driverle

*Louis C.K. net*

## PRAISE FOR TRANSCENDENCE

"Sure-footed guides through the treacherous landscapes of infinite possibility, R.U. Sirius and Jay Cornell are lucid, witty, and stealthily profound. *Transcendence* is a refreshingly pro-human report on transhumanism and the 'Rapture of the Nerds.'"

—Douglas Rushkoff, author of *Present Shock: When Everything Happens Now*

"R.U. Sirius presents us with his own psychedelic guide to the galaxy in this adventurous idea-rich book, bootstrapping on emerging technologies that beckon us to take control of our evolutionary destiny and lead humanity towards radical new landscapes of mind, of dream, of cosmos, of possibility. Metaphors for self-transcendence are 'literalized' as we 'timelapse' our capacity to artfully remake the world and ourselves."

—Jason Silva, host of Brain Games and Shots of Awe

# TRANSCENDENCE

# TRANSCENDENCE

THE DISINFORMATION ENCYCLOPEDIA OF

# TRANSHUMANISM

AND THE

# SINGULARITY

R.U. SIRIUS

JAY CORNELL

This edition first published in 2015 by Disinformation Books, an imprint of
Red Wheel/Weiser, LLC
With offices at:
665 Third Street, Suite 400
San Francisco, CA 94107
www.redwheelweiser.com

ISBN: 978-1-938875-09-0

Library of Congress Cataloging-in-Publication Data available upon request.

Cover design by Jim Warner
Interior by Deborah Dutton
Typeset in Adobe Caslon Pro, ITC Franklin Gothic, ChaletComprime Cologne
Sixty and OCR A Std.

Printed in the United States of America.
EBM
10  9  8  7  6  5  4  3  2  1

*R.U. Sirius would like to dedicate this book to his love, Eve Berni, to better and better health.*

*Jay Cornell would like to dedicate this book to his parents, without whom he would not have been possible, and to his late friend Chuck Frutchey, who would have thought it was cool.*

# CONTENTS

TRANSCENDENCE

# ACKNOWLEDGMENTS

Special thanks are due to Ben Goertzel of Humanity+, who did all sorts of stuff for us; Greg Campbell, a.k.a. Surfdaddy Orca (RIP); and Greg's son MacGregory Campbell. Also thanks to all the folks we quoted, but particularly James Kent, Hank Pellissier, and, again, Ben Goertzel, for their massive contributions.

Also, thanks to Steven Kotler, James Kent, J. Storrs Hall, Chris Grayson, Erik Sayle, Franco Cortese, Gregory Benford, Matt Householder, and John Bartelt, for reviewing some of this material for us . . . but don't hold any of them responsible for that awful error or omission you think you found on page whateverthefuck.

# INTRODUCTIONS

# THE COMING TRANSHUMAN: BEST. THING. EVER.

## BY R.U. SIRIUS

There have been some astonishing events during the lifetime of your authors. The Beatles on *The Ed Sullivan Show*. The passage of civil rights legislation in the United States. The publication of *Fear and Loathing in Las Vegas*. The rise of the Internet. The fall of the Berlin Wall. Designer drugs. Watson's victory on *Jeopardy*. And what can we say about Scarlett Johansson?

But thus far, there's nothing that has happened in our lives—indeed, nothing that has *ever* happened in the lives of any human beings throughout the entire *Homo sapiens* experience—that can come close to what many serious and reputable scientists and technologists predict will happen during the 21st century. We will, some say, be designing our own evolution, making ourselves into different creatures—radically enhanced versions of human beings. We are, they say, *transhumans*—humans transitioning, via technology, into something grander . . . the posthuman.

Transhumanism is the international movement that advocates this self-directed evolution. It proposes that we use science and technology to overcome the "natural" limitations experienced by humanity. As stated in the first point of the eight-point Transhumanist Declaration: "We envision the possibility of broadening human potential by overcoming aging, cognitive shortcomings, involuntary suffering, and our confinement to planet Earth."

Beyond ending aging, gaining greater intelligence, and conquering the stars, many transhumanists look forward to a plethora of other alterations to the ordinary human condition:

- The Singularity—the creation of machine intelligences that exceed the capacities of our biological brains.

- Radical robotics and the end of economic scarcity and boring labor.

- The ability to replicate individual minds and put them into solid-state bodies or virtual environments.

- Improved physiological strength, sexual pleasure, and the intentional mutation of the human body.

- The ever-increasing interconnectedness and empowerment of humans via ever-faster and more powerful communications technology and, with it, the potential for vastly more productive and creative group minds.

- Individual control over mental and emotional states for enhancing functionalities and ecstasies.

In other words, transhumanist hopes and expectations seem science fictional in comparison to the quotidian (if not desperate) lives most of us humans currently lead.

If it does occur, this radical shift in the human situation will be largely due to technological progress in nanotechnology, biotechnology, neuroscience, and artificial intelligence. (This combination is often abbreviated to nano-bio-info-cogno or NBIC.) Many transhumanists passionately believe that these technologies will even bring immortality.

Some wags have called this vision "the rapture of the nerds." The transhuman future does indeed seem to make many of the same promises as most religions: an immortality in a sort of heavenly place (a technologically improved world, real or digital, where most or all of the ordinary difficulties of biological life are transcended).

But even these ultra techno-optimists recognize that these technologies don't come without risks. Professor Nick Bostrom, for example, an early transhumanist and a professor at the James Martin 21st Century School at Oxford University, suggests that some of the biggest existential risks facing the human race come from expected future developments in NBIC: human-like terminator robots, molecular nano weapons, new genetic strains of viruses, thought control mind interfaces—to name a few. We've all seen the movies.

It's a complicated dilemma. Our species—now over six billion strong and growing—is having a difficult time muddling through with our limited capacities to know and use our resources in sustainable ways. We will likely need some advanced NBIC capacities to realize humanity's *Star Trek* potential to alleviate human suffering globally, fix the environment, evolve smarter and more empathetic selves and societies, and reach the stars. But can we keep technology from making things worse instead of better? Will we be ruled by unsympathetic robot overlords?

Some technology luminaries, such as Microsoft researcher Gordon Bell, acknowledge that the possibility of transhuman technological change is at hand, but feel that the human population will destroy itself before it happens. Others (including Intel founder Gordon Moore) are skeptical that humanity will ever gain the more extreme and transformative versions of artificial intelligence, nanotechnology, and so forth predicted by tech optimists. Moore's skepticism is particularly poignant, since his famous "law" is the basis for the prediction that technological innovation is accelerating, and that the next twenty to fifty years may yield not only radical technological advances, but perhaps even The Singularity of hyperintelligent machines (see Moore's Law).

Whether or not you view the transhumanist movement as a wild and wacky expression of the madness of late capitalism—or you think it is scientifically and philosophically on sound footing—this book takes you on a trip. It's your comprehensive A–Z guide for becoming conversant in the controversial sciences, ideas, and cultures of transhumanism. It provides a multilayered look at the science and philosophy behind the accelerating advances in artificial intelligence, nanotechnology, neuroscience, synthetic biology, robotics, and so many other advances that are leading toward this hypothetical transhuman future. This book also gives you a juicy insider's view of transhumanist subcultures, bringing you up to date on the personalities and influential institutions (like Google, Singularity University, and DARPA) that are participating in transhumanist technological efforts. It is also a handy reference book for those who are already hooked in to the transhumanist vision.

Indeed, once you have finished this book, you will be able to make scintillating party chatter every time a friend or family member starts showing off his or her latest gadget purchase: "Yeah. You think *that's* something. Well, in 2045 . . ."

So prepare for acceleration. Fasten your safety belts, bring your brain into an upright position, order your favorite nutraceutical refreshment, and *read this book.*

# TECHNO-OPTIMISM: A BRIEF HISTORY

## BY JAY CORNELL

### "PROGRESS" IS BORN

Only in the 18th century Enlightenment did the concept of progress become widespread. Earlier, most people thought of history in terms of a fall from a past Golden Age, or perhaps repeating cycles. (If they thought of such things at all. Mostly they just worried about their next meals.)

With the Industrial Revolution, progress became almost synonymous with science and technology. By the late 19th and early 20th century, we see the beginnings of modern science fiction (Verne, Wells), and prototypes of today's hackers and geeks (Edison, Tesla, Tom Swift). Tellingly, we also see early instances of techno-optimistic wishful thinking: the telegraph, dynamite, and airplanes (and later, movies and television) were all heralded, sometimes even by their inventors, as tools that would end war. The First World War was a huge setback for all optimists, but the techno-optimistic spirit soon recovered.

### SCIENCE FICTION IS BORN, SCIENCE FACT TRIUMPHS, AND THEN GETS SCARIER

Radio, aviation, medicine, and much more were changing the world, so the prophets of science got more attention. Writer, editor, and radio and television pioneer Hugo Gernsback was that era's premier techno-optimist, publishing magazines such as *Modern Electrics* (1908) and *The Electrical Experimenter* (1913). His science fiction novel *Ralph 124C 41+* (1911) included television, radar, solar energy, synthetic food, space travel, and lots more. Even the title is in proto-textspeak: Ralph's last name means "one to foresee for one another." In 1926 he launched *Amazing*

*Stories*, the first magazine dedicated to science fiction. (Heard of the Hugo Awards? That's our Hugo.) Starting in 1938, John W. Campbell, Jr.'s *Astounding Science Fiction* took the lead, nurturing such greats as Asimov and Heinlein and ushering in the Golden Age of science fiction.

For many, Soviet "scientific socialism" promised a more efficient and productive future, as did eugenics and (briefly) the Technocracy movement. Science and technology were still largely participatory: if you had an idea for a better automobile or airplane, or wanted a television, you built one. By 1945, inventions like antibiotics, computers, radar, and the atomic bomb had helped win World War II, making it seem closed-minded to dismiss "that Buck Rogers stuff," but the future looked darker.

## FROM UTOPIAS TO DYSTOPIAS

After Hitler and Stalin, eugenics and scientific socialism lost their luster. The threats of nuclear war, Sputnik, and pollution made interwar techno-optimism seem naïve. Science fiction continued to flourish, but "post-apocalypse" was no longer a purely theological concept. Even optimistic futures began to seem plastic and bureaucratic. Science had become Big Science, done in expensive laboratories with mainframes and other tools that only organizations could afford. Progress now came from conformist Organization Men. Lone inventors and independent scientists were obsolete.

## THE ZEITGEIST SHIFTS AGAIN

The 1960s–70s saw environmentalism and various anti-technology ideas gain strength, but techno-optimism was reborn in a new, post-psychedelic form. It was manifested in the *Whole Earth Catalog*, life extension research, Timothy Leary's turn toward technology, and more. IBM and HP didn't care about personal computers, so hackers built their own. NASA was too staid, so the L5 Society was founded to kickstart the creation of one of Gerard K. O'Neill's space colonies. The future regained a blue-sky, do-it-yourself, distributed, anti-bureaucratic feeling. Beyond new inventions, it was about expanding human potential using all available techniques (legal or illegal). Gernsbackian hands-on techno-optimism had dropped acid and returned with a bunch of new ideas.

## TODAY

Transhumanism is today's edgiest form of techno-optimism. It's similar to earlier expressions, with Tesla-like eccentric scientists, modern-day Edisons pushing the boundaries of electronics, new versions of Jobs and Wozniak building transformative tech in garages, and lots of cheerleaders and fans, but there's more. Accelerating advances in computing and biotechnology promise (and threaten) to create not just better objects and a transformed society, but transformed human beings: healthier, smarter, stronger, and with "superhuman" abilities. This focus on improved humans, and the willingness to actually try it (often via self-experimentation), is central.

The other core aspect is decentralization. True, there will be no Singularity without Big Science and Big Business: nobody can make billions of smartphones in a garage. But much of the real action is in using new knowledge and new tools in personalized, homemade, open-sourced, self-funded and crowdfunded ways.

## NEW TOOLS VS. OLD RULES

Along with all this comes a conflict we've seen before: new tools (or old tools, made affordable) not covered by old rules. Much of transhumanism is happening without central planning or permission from politicians, CEOs, or lawyers, so Authorities Are Concerned. Their worries aren't all easily dismissible, but it's critical that the vested interests, naysayers, pearl-clutchers, and tear-squeezers not have the final word and stop progress with red tape.

## YOUR FUTURE: PROMISING, DECENTRALIZED, DANGEROUS

Tesla, Edison, and Gernsback would be thrilled at today's discoveries and inventions. Looking at the renaissance of "makers," at the giddy enthusiasm of regular people pursuing scientific dreams, they couldn't help but approve. Sure, they'd have concerns, especially about the more extreme proposals for human transformation, but I'm sure that Gernsback (at least), given the chance, would have uploaded his mind to a computer.

Whether or not you believe the predictions, whether you fear all this or want to help it happen (and you can!), transhumanism is what some of today's best minds are working on and arguing about. It's big, and happening ever-faster. Welcome to the age of *Tom Swift and His Homemade Biomedical Implant*. (And, hopefully, not to the age of *Tom Swift and His World-Devouring Nanobots*.)

One way or another, advanced science and affordable, powerful technology will keep changing the world, and humanity with it. Get ready for a wild ride.

The contents of this book combine the words of
your authors with words from other writers and
experts in these fields as they were written
for the Transhumanist webzine, h+—the official
website for the Humanity Plus organization,
and for Acceler8or. The author R.U. Sirius was
editor of both sites.

# ABOLITIONISM

*(See also Designer Babies)*

This may be crazy, but what say we end suffering in all sentient beings? David Pearce, an English transhumanist, proposes abolitionism: a new movement to completely end all such sentient misery. And you thought Europeans weren't ambitious anymore.

The primary methodology would be germline engineering of "designer babies" for maximum happiness and minimum pain, in combination with various other advanced technologies for ending scarcity and eliminating physical vulnerabilities in humans.

## PEACE AMONG THE ANIMALS

But it doesn't stop there. Pearce, an animal rights enthusiast, figures, *Hey, let's make the animals peaceful and happy!* "Gene therapy will be targeted both on somatic cells and, with even greater forethought, the germline. If cunningly applied, a combination of the cellular enlargement of the mesolimbic dopamine system, selectively enhanced metabolic function of key intra-cellular sub-types of opioidergic and serotonergic pathways, and the disablement of several countervailing inhibitory feedback processes will put in place the biomolecular architecture for a major transition in human evolution." Quite a mouthful, but if you're trying to end all suffering on Earth, it figures to be a bit complicated.

**James Kent:** Pearce's intellectual embrace of paradise engineering places him on the cusp of a modern philosophical movement that eschews Darwinian fatalism and looks to a post-Darwinian future where humans are freed from the cynical bonds of genetic expression and natural selection.

TRANSCENDENCE

**David Pearce:** Here are some grounds for cautious optimism that transhumanism can eliminate suffering in humans:

### 1) We Shall Soon Be Able to Choose Our Own Level of Pain Sensitivity

A revolution in reproductive medicine is imminent. Clearly, our emotional response to raw pain is modulated by the products of other genes. But recent research suggests that variants of the SCN9A gene [see "Pain's in the Genes," *Science Magazine*, March 8, 2010, or take Pearce's word for it] hold the master key. Thus, in a decade or two, preimplantation diagnosis should allow responsible prospective parents to choose which of the SCN9A alleles they want for their future children—leading in turn to severe selection pressure against the SCN9A gene's nastier variants.

At present, we can't envisage safely choosing one of the (extremely rare) nonsense mutations of SCN9A that eliminates physical pain altogether. A future world of nociception without any phenomenal pain at all will depend on advances in neuroprosthetics and artificial intelligence. Yet by selecting benign alleles of SCN9A both for ourselves and our children, the burden of suffering can soon be dramatically diminished.

### 2) We Can Soon Choose How Rewarding We Want Our Daily Life to Be

The brain is a dauntingly complex organ. Yet the biological roots of mood and emotion are primitive and neurologically ancient. Their metabolic pathways are strongly conserved in the vertebrate line and beyond. A research paper published in *Nature* in 2007 illustrates how the presence or absence of a single allele may dramatically enrich or impair the quality of one's entire life.

How can we guard against unanticipated side effects from novel gene therapies? What will be the inevitable "unintended consequences" of life-changing innovation? And what will be the societal implications of a population biologically predisposed to enjoy richer and happier lives?

Genetic tweaking to promote richer experience is just a foretaste of posthuman sentience. I predict that our descendants will enjoy gradients of genetically preprogrammed bliss every day of their lives.

# ARTIFICIAL GENERAL INTELLIGENCE

*(See also Artificial Intelligence, The Singularity)*

The original goal of the AI field was the construction of "thinking machines"—that is, computer systems with human-like general intelligence. Due to the difficulty of this task, for the last few decades the majority of AI researchers have focused on what has been called "narrow AI"—the production of AI systems displaying intelligence regarding specific, highly constrained tasks. Common narrow AI applications include the AI linguistics underlying Google and other search engines; the AI planning and scheduling software used throughout the military and industry; the AI fraud detection software underlying modern credit card operations; the AI used by those friendly folks at the NSA to find patterns in your phone calls; and the AI gaming software underlying everything from IBM's chess-playing Deep Blue to the bots in massively multiplayer online games. These are all amazing achievements (albeit the uses can be questionable) that have a common narrowness of scope, which is why Ray Kurzweil has characterized them as "narrow AI."

In recent years, however, more and more researchers have recognized the necessity and feasibility of returning to the original goals of the field. Increasingly, there is a call for a transition back to confronting the more difficult issues of "human level intelligence" and more broadly "artificial general intelligence (AGI)."

AGI describes research that aims to create machines capable of general intelligent action. The term was introduced in 2003 in order to avoid the perception that the field was only about creating human-level or human-like intelligences, which is also covered by the term "Strong AI." AGI allows for the inclusion of nonhuman as well as human models of general intelligence.

One approach to Artificial General Intelligence may involve reverse engineering the brain. This quest is described in detail in Jeff Hawkins' book *On Intelligence*.

*Written with Surfdaddy Orca*

## When Will We Get Human-Level AIs?

**Ben Goertzel, Seth Baum, Ted Goertzel:** When will we have Artificial General Intelligences (AGIs) we can talk to? Ones as smart as we are, or smarter?

The majority of the experts who participated in a 2010 AGI survey by *h+* were optimistic about AGI coming fairly quickly, although a few were more pessimistic about the timing. However, all the experts in the study, even the most pessimistic ones, gave at least a 10 percent chance of some AGI milestones being achieved within a few decades. The experts were asked to give estimates on each of four milestones:

- Passing the Turing test by carrying on a conversation well enough to pass as a human.
- Solving problems as well as a third grade elementary school student.
- Performing Nobel-quality scientific work.
- Going beyond the human level to superhuman intelligence.

There was consensus that the superhuman milestone would be achieved either last or at the same time as other milestones. However, there was significant divergence regarding the order of the other three milestones. One expert argued that the Nobel milestone would be easier than the Turing Test milestone precisely because it is more sophisticated: to pass the Turing test, an AI must "skillfully hide such mental superiorities." Another argued that an AI that passes the Turing test needs the same types of intelligence as a Nobel AI "but additionally needs to fake a lot of human idiosyncrasies (irrationality, imperfection, emotions)." Finally, one expert noted that the third grade AI might come first because passing a third grade exam might be achieved "by advances in natural language processing, without actually creating an AI as intelligent as a third-grade child." This diversity of views on milestone order suggests a rich, multidimensional understanding of intelligence. It may be that a range of milestone orderings is possible, depending on how AI development proceeds.

One observed that "making an AGI capable of doing powerful and creative thinking is probably easier than making one that imitates the many, complex behaviors of a human mind—many of which would actually be hindrances when it comes to creating Nobel-quality science." He observed, "humans tend to have minds that bore easily, wander away from a given mental task, and that care about things such as sexual attraction, all of which would probably impede scientific ability, rather than promote it." To successfully emulate a human, a computer might have to disguise many of its abilities, masquerading as being less intelligent—in certain ways—than it actually was. There is no compelling reason to spend time and money developing this capacity in a computer.

# ARTIFICIAL HIPPOCAMPUS

*(See also Implants, Prosthetics, Neurotechnology)*

In the part of our brain called the hippocampus, our short-term memories become long-term memories through a process involving the movement of electrical impulses through neurons. A number of projects have looked for ways that an injured—or maybe just insufficient—hippocampus can be bypassed via the implantation of miniaturized electronic equipment.

In a set of experiments announced in 2013, neuroscientists from the University of Southern California (USC), Wake Forest University (WFU), the University of Kentucky, and Defense Advanced Research Projects Agency (DARPA) have been able to use a device to record patterns in tissues of mice and monkeys related to specific memories, and then replicate those by stimulating healthy brain cells with electricity—making up for functions lost to damage in other brain cells.

Researchers are predicting that humans could be using such devices to replace lost memories in five to ten years. The FDA may be willing to hurry this along, given that the immediate application will be for Alzheimer's, an ever-growing crisis that will likely be even more critical by that point.

Projecting a little bit further out, artificial hippocampus devices are a giant step toward the transhumanist dream of devices that can record and preserve all our memories.

**David Pescovitz:** Biomedical engineer Theodore Berger at the University of Southern California in Los Angeles has developed an artificial hippocampus. To do this, Berger built mathematical models of neuronal activity in a rat's hippocampus and then designed circuits that mimic those activities. Joel Davis at the Office of Naval Research, a sponsor of Berger's work, said, "Using implantables to enhance competency is down the road. It's just a matter of time." While Berger's work is a far cry from a hard drive for the brain, I'm intrigued by the notion of being able to "back up" my memory just in case.

# ARTIFICIAL INTELLIGENCE

*(See also Neurobotics, Robotics)*

As opposed to AGI (Artificial General Intelligence), which doesn't really exist yet, Artificial Intelligence (instruments that can autonomously perform specific constrained tasks) is all around us. In addition to search engines, credit card fraud detection, snooping on phone calls, and gaming, AIs are answering our phones and giving customer assistance (badly!), beating humans at chess and *Jeopardy*, driving cars, targeting drone strikes, doing most of the flying that we attribute to our airline pilots, and collecting data and doing rudimentary analyses of conditions in space. It's even correcting things I type here, although sometimes I wish it would stop.

As we all know from dealing with automated phone systems, one of the limitations of constrained AI is that real life situations often don't correspond to a fixed set of options. The distinction between current AI and AGI (or humans or even animals) is its minimal ability to respond to unexpected stimuli.

Transhumanists are most enthused about developments in contemporary AI that appear outwardly to mimic human abilities like learning and showing emotions.

**Steven Kotler:** UC Irvine neuroscientist Jeff Krichmar's AI-based bots develop personalities because, instead of preprogramming behaviors, these robots have neuromodulatory systems or value judgment systems (move towards something good, move away from something bad) that are modeled around the human's dopaminergic system (for wanting, or reward-based behaviors) and the noradrenergic system (for vigilance and surprise). When something salient occurs—in the case of Krichmar's bots, that's usually bumping into a sensor in a maze—a signal is sent to its brain telling the bot to react to the event and remember the context for later. This is conditional learning and it mimics what occurs in real brains.

# ARTIFICIAL LIFE

*(For artificial biological life, see Synthetic Biology)*

## EVOLUTION IN DIGITAL FORM

How can something as complex as life arise without a Creator? How can mere humans even hope to replicate it? Well, it turns out that it's not as difficult as you might think. For many years artificial life (or ALife) researchers have been simulating biological life using computers (or with robotics or biochemistry). Right now, you could be playing God in the simplified world of a browser window.

## DOTS ON A GRID

Stanislaw Ulam and John von Neumann first discussed cellular automata in the 1940s, but the field really took off when John Conway unveiled his elegant Game of Life in 1970, and there are now many free versions online. They all use a simple grid and simple rules. Each square either contains a dot ("alive") or is empty ("dead"). A dot stays alive if it has two or three neighbors, or dies if it's too crowded (four or more neighbors) or too lonely (one or no neighbors). When an empty square borders three dots, a new dot is "born" there. Now, start with a bunch of dots and watch the fun.

## LET THERE BE ALIFE

Surprising patterns will emerge. Still lives (such as a square block of four dots) will stay static. Oscillators (the simplest is three dots in a row) will cycle between two or more states. Gliders and other spaceships will fly across the grid. Glider guns will emit streams of gliders. Puffer trains will crawl, leaving persistent smoke. Collisions will create more patterns. Moving, eating, excreting, birth, death—all visible in dots on your screen.

## OPENWORM

The open source OpenWorm project has produced the most advanced digital life form so far. The nematode *Caenorhabditis elegans*, with all of its one thousand cells and the neurons connecting them, is modeled entirely in software. The latest version wriggles realistically and at the same speed as a real one, though this is due to hard-coded instructions, not the 302-neuron brain, which is still being developed.

## SO WHY DOES ALIFE MATTER TO TRANSHUMANISTS?

Several reasons. The Game of Life is "Turing complete": it can simulate any real-world computer or computer language. It's used in cryptography. As a model it's helpful for anyone researching synthetic biology. Some physicists even believe that our universe is best described as a cellular automaton.

But perhaps the core lesson is that the complex (and solutions to complex problems) need not be designed from the top down: astounding levels of complexity can arise from simple rules plus randomness. Remember that the next time someone claims that big solutions require big bureaucracies, or when someone scoffs at the empowering possibilities of transhumanist tech.

# AUBREY DE GREY

*(See also Longevity/Immortality, The Methuselarity,
SENS Research Foundation)*

Somewhere around the turn of the millennium, Aubrey de Grey, an English biogerontologist who is now as famous for his Father Time beard as he is for his outspoken vision of radical life extension, looked at aging as an engineering problem and decided: *Eureka! We can do this.*

Since then, de Grey has become the preeminent spokesperson for the scientific quest to end death. He formed the Methuselah Foundation, which offers a cash prize for a particular longevity breakthrough, and the SENS (Strategies for Engineered Negligible Senescence) Research Foundation, which is conducting scientific research on longevity, using principles de Grey laid out in his 2007 book, *Ending Aging.* He has appeared on *60 Minutes, The Colbert Report,* and the Barbara Walters special report "Live to 150, Can You Do It?" and recently wrote a guest opinion piece as part of a 2013 *Time* magazine cover story on Google's recently announced project to conquer death.

## THE REAL DE GREY

De Grey is an easygoing fellow known for his fondness for beer and for generally not adhering to the intense physical regimens followed by many other longevity fanatics. Having ridden in the back seat of his car one time, I can report that de Grey likes depressing British post-punk, which totally humanizes him as far as I'm concerned.

## Aubrey's Approach

**Michael Anissimov:** Instead of exclusively studying the complex biochemical processes of aging in detail, as in gerontology, or ameliorating the worst symptoms of age-related decline, as in geriatrics, de Grey and his supporters advocate an "engineering approach" to aging that asks, what are the main categories of age-related biochemical damage, and how can we fix them? The idea is not to eliminate the sources of age-related damage, but to fix the damage fast enough so it doesn't accumulate and cause health problems. This is far easier than deciphering all the intricacies of the biochemistry of aging . . .

## It's Not Personal

**Aubrey de Grey (interviewed by R.U. Sirius):** I'm actually not mainly driven by a desire to live a long time. I accept that when I'm even a hundred years old, let alone older, I may have less enthusiasm for life than I have today. Therefore, what drives me is to put myself (with luck) and others (lots and lots of others) in a position to make that choice, rather than having the choice progressively ripped away from me or them by declining health. Whether the choice to live longer is actually made is not the point for me.

# AUGMENTED REALITY

*(See also Sousveillance)*

## COMPUTER: AUGMENT!

Augmented reality (AR) means overlaying digital information on the real world, live and in realtime. Think of the game score at the bottom of the TV screen. In more advanced forms, the information is deeper and interactive, and might not just be on a screen, but built into your glasses, or contact lenses, or fed directly into your eye or even your brain.

Here's how AR can work today, with just a smartphone or tablet. You turn on the camera and aim it at a building. The GPS knows where you are, the accelerometer and gyroscope know where you are pointing, and the AR software does its thing, recognizing the building and doing some research. Links appear on the screen: the history of the building, property tax records, the current owners, current tenants, the architect's blueprints, newspaper mentions. They want *that much* for 4C, the empty two bedroom unit on the north side? It's been vacant for three months! Maybe they'll take an offer.

## ENDLESS POSSIBILITIES

Let's just say: education, medicine, architecture and construction, art, industrial design, military, gaming and entertainment, navigation, marketing, tourism, translation, and lots more.

AR has tremendous potential as assistive devices for the impaired. The autistic could get help with reading facial expressions. Maybe you can't see well at night, but the camera in your glasses can. Grandma can't remember to take her medicine, so her glasses remind her every day, and they can tell when she's got the right bottle and the right number of pills.

## SO WHAT'S NOT TO LIKE? LOTS

The significant parts of physical reality we call "other people" aren't always happy being on camera, much less being video- and audio-recorded and perhaps facially-recognized and Googled all at the same time. Already Google Glass wearers are nicknamed "glassholes" and banned from strip clubs. People who sell tickets to experiences like movies, plays, concerts, and sporting events don't want you recording or streaming your experience. Casinos worry about augmented cheating.

Safety is a concern. Yes, wearable tech like Google Glass can function as a hands-free smartphone, which is safer than holding one while driving, but how do you know those other drivers aren't watching porn, or trying to find the Facebook page of that hot-looking driver one lane over? Already one person has been ticketed for driving while augmenting, though she claims her device was switched off at the time and the citation was dismissed.

Hospitals worry about patient privacy, and privacy concerns in general are linked to larger concerns about liberty. The grassroots group Stop the Cyborgs is encouraging people to ban Google Glass and other wearable sousveillance tech from their property. They're concerned about a future in which privacy is impossible, not because of an overt 1984-style Big Brother dictatorship, but due to an "'iron cage' of surveillance, calculation, and control."

## ULTIMATE AUGMENTATION

Imagine a set of wearable sensors combined with a neural interface. Anything you saw or heard (and tasted, felt, and thought) could be augmented with digital data. Your inner experience could be recorded, à la the movie *Brainstorm*. The potential is incredible. However, there will likely be problems when the boundaries of the physical and digital worlds get blurry, and having all that experience in digital form would make a tempting target. You would not want government snoops or random hackers accessing the digital overlay of your life, or even worse, manipulating it. Maybe those Stop the Cyborgs people have a point.

## Enhanced Vision

**Michael Anissimov:** A key application of augmented reality that has only just begun to be probed is virtual "X-ray vision." Using digital eyewear like the Vuzix Wrap 920AV, revealed at 2009 CES in Las Vegas, engineers and maintenance technicians might be able to "look inside" machines without ever taking them apart, allowing for better analysis of complex systems and faster repair. Combined with microsensors, augmented reality could provide an intuitive way to analyze the features of systems that are not immediately obvious through visual scrutiny alone. In the further future, augmented reality instruction manuals could replace paper manuals entirely, giving the user all the information they need without even requiring them to take their eyes off the product . . .

## Augmented Reality in Science Fiction

**Michael Anissimov:** One of the most extensive fictional portrayals of augmented reality appears in Vernor Vinge's book *Rainbows End* (2006). The world in the book includes a ubiquitous augmented reality that has replaced conventional screens as the primary medium for accessing the Internet and communicating with others remotely. The characters in the story use contact lenses to view augmented reality, removing them only to sleep. By combining augmented reality graphic overlays with haptic feedback and robots, the story presents a world where the line between the virtual and the real is thoroughly blurred. Numerous AR worlds are available for any user, but the most popular are built in collaborative units called "belief circles." In the novel, entire parks are used exclusively as augmented reality playgrounds.

# BIOHACKING

*(See also Citizen Scientists, Grinders)*

## GARAGE BIOLOGY

In the 1970s, computers got cheap enough that individuals could afford them. Enthusiasts were no longer constrained by what the school or the boss would allow (or by what they could get away with), and could do what they pleased at home. Computer hacker culture took off, sparking the personal computer revolution that changed (and continues to change) the world.

Biology is now starting on that same trajectory. Cheaper equipment, open source ideology, a homebrew DIY ethic, and impatience with the conservatism and bureaucracy of mainstream science are all fueling a diverse group of people called biohackers (a.k.a. biopunks and DIYbiologists).

## DIY TOOLS

Biohackers often build their own equipment. Sometimes the plans are simply released as open source hardware, like the DremelFuge, a 3D-printable attachment for rotary tools that allows them to work like a centrifuge costing hundreds of dollars. Other times the makers also sell their creations as kits or fully assembled, like the SpikerBox, a "bioamplifier" that allows citizen neuroscientists to see and hear the action potentials of living neurons in invertebrates. The SpikerBox has been so successful that even "real" neuroscience labs are using it.

## BIOHACKING YOGURT AND YOURSELF

DIYbiology doyenne Meredith Patterson, creator of glow-in-the-dark yogurt, has also made steps toward a cheap melaminometer for detecting the melamine used to adulterate milk and eggs in China. The idea is to add jellyfish genes to yogurt bacteria. However, the project is currently on hold because synthetic biology labs can't yet engineer a suitable detector. Other biohackers hope to make glowing plants to replace street lamps.

Biohackers often self-experiment. Tim Ferriss brought self-biohacking to the mainstream with his bestseller *The 4-Hour Body*. He described his experiments in carefully tracking diet and exercise to improve his physical and mental health, with the aid of the inexpensive DNA analysis services of 23andMe. (For more, see Quantified Self.)

Self-biohacking can also go beyond what can seem like A+ science fair projects. For the more extreme forms, see Grinders.

## HEY GANG, LET'S MAKE A GMO

Biohackers have a growing support network, with groups like the Registry of Standard Biological Parts, the OpenWetWare community, and the BioBricks Foundation supporting them. There are now dozens of community biolabs, like Brooklyn's Genspace and BioCurious in Sunnyvale, throughout the world.

## BIO-TERROR OR BIO-ERROR

You may want to be Doctor Moreau on your own island, but many people don't want you to be. The biological anarchy of individuals making their own GMOs brings fears of intentional or accidental releases of modified organisms, and your home lab can look like a terrorist operation to overzealous police. But the genie is out of the bottle, and biohacking is on track to change the coming decades in ways we can't yet imagine.

## How Meredith Patterson, the Doyenne of DIYbio, Got Started

**Meredith Patterson (interviewed by Tyson Anderson):** As an intro to a talk, I isolated chickpea DNA using non-iodized salt, shampoo, meat tenderizer, and a salad-spinner for a centrifuge, and that really blew people away. So that was the point when I started spending my free time reading old papers and thinking more about how to do more advanced genetics research at home.

## Is DIYbio Dangerous?

**Meredith Patterson (interviewed by Anderson):** The chances of someone accidentally creating a dangerous organism and the chances of it surviving in the environment outside a laboratory are vanishingly low.

## Mainstream Support

**Parijata Mackey:** Mainstream science is increasingly friendly to DIYbio. DIY-biologist Jason Bobe works on George Church's Personal Genome Project (PGP), which shares and supports DIYbio's drive to make human genome data available for anyone to use.

# BRAIN-BUILDING PROJECTS

*(See also Artificial General Intelligence, Neurotechnology)*

## HENRY MARKRAM'S BLUE BRAIN PROJECT

This is the first comprehensive attempt to reverse-engineer the mammalian brain. The brain processes information by sending electrical signals between neurons using the "wiring" of dendrites and axons. In the cortex, neurons are organized into basic functional units—cylindrical volumes—each containing about ten thousand neurons that are connected in an intricate but consistent way. These microcircuits, known as neocortical columns, are repeated millions of times across the cortex.

## BUILDING AN ACCURATE MICROCIRCUIT NEURON

The first step of the project is to re-create this microcircuit, down to the level of individual neurons. The microcircuits can then be used to simulate such things as variation in particular neurotransmitters, mimicking what happens when the molecular environment is altered using drugs.

The project believes that they will have created the cellular equivalent of a human brain by 2023.

## SPAUN BRAIN SIMULATION PROJECT

The Spaun Brain Simulation Project, out of the University of Waterloo, Canada, claims to be the world's largest functional brain model. The model consists of about 2.5 million spiking neurons. It also has a single eye and an arm, which may be part of the team's claim of functionality.

## IBM'S BLUE MATTER PROJECT

IBM's Dharmendra Modha has a vision of cognitive computing. "Cognitive computing seeks to engineer the mind by reverse engineering the brain," says Modha, a researcher at IBM's Almaden Research Center, just south of San Francisco. "The mind arises from the brain, which is made up of billions of neurons that are linked by an Internet-like network."

## BLUE MATTER'S KITTEH BRAIN

Modha's vision may have taken a giant leap forward with a 2009 announcement, at a high-performance computing conference in Portland, Oregon, of a joint IBM project led by Modha with researchers from five universities and the Lawrence Berkeley National Laboratory. Dubbed "Blue Matter," a software platform for neuroscience modeling, it pulls together archived magnetic resonance imaging (MRI) scan data and assembles it on a Blue Gene/P Supercomputer. IBM has essentially simulated a brain with one billion neurons and ten trillion synapses—one they claim is about the equivalent of a cat's cortex, or 4.5 percent of a human brain.

The funding for the project comes from Phase 1 of the US DARPA SyNAPSE project that seeks "to discover, demonstrate, and deliver algorithms of the brain via a combination of (computational) neuroscience, supercomputing, and nanotechnology." IBM's announcement signals significant progress toward creating Modha's future computer, one that will simulate and emulate the brain's abilities for sensation, perception, action, interaction, and cognition, and rivaling the brain's low power, energy consumption, and compact size.

This starts to raise interesting questions, as several bloggers noted, somewhat tongue-in-cheek, after the announcement. "So, will Blue Gene get the sudden urge to lick itself?" asks one blogger. "I find myself feeling sorry for the virtual 'cat,'" says another. "If you create an animal-brain-based system that's capable of learning from experience, feeling basic emotions, then shouldn't it have the same rights as a flesh-and-blood pet?"

Nothing has been heard publicly from this project since 2009, but that's the way IBM works.

*Written with Surfdaddy Orca*

## Blue Brain in a VR Agent

**Surfdaddy Orca:** Project Director Henry Markram told the recent Science Beyond Fiction conference in Europe that Blue Brain's "neocortical column" is being integrated into a virtual reality agent—"a simulated animal in a simulated environment, so that the researchers will be able to observe the detailed activities in the column as the animal moves around the space."

Markram believes that, "building up from one neocortical column to the entire neocortex, the ethereal 'emergent properties' that characterize human thought will, step-by-step, make themselves apparent."

# ODY SCULPTING

*(See also Stem Cells)*

The popular notion of body sculpting involves physical exercise and diet regimens suited to the individual so that one might have the body one desires. This may be combined with some type of more common body modification surgery, most frequently liposuction, to remove fatty deposits.

## NEVERMIND THE GYM, I WANNA BE A CYBORG

But of course, those closer to the transhuman edge have some different ideas. The neo-cyberpunk Shadowslang glossary says body sculpting is "the process of having the body totally remolded by vat-grown tissues and advanced plastic surgery," which I think is partly speculative, although I can't swear to it.

Transhumanist dreams of body sculpting tend to involve genetic and cyborgian techniques that may, on the one hand, give the recipient a conventionally impressive body or, on the other hand, may involve something more imaginative, including being able to switch genders at will. Or it could describe completely imaginary bodies in virtual spaces that, in fact, may become the major locus for our lives.

## Bigger Penises, of Course

**Valkyrie Ice:** Our increasing knowledge of genetics is going to make things like customized body sculpting not only easier, but probably pretty cheap too. We can print organs now, and use stem cells to cosmetically alter our bodies. They've even tested building rabbit penises. Seriously, how long do you think it's going to be before an organ printer is routinely used to create larger penises? And how long after that do you think it could be used to create a customized elf ear? Or even a prehensile tail for a human being? Once we've cracked the DNA code for it, we can program a stem cell to become any other kind of cell.

## Or Tails, Perhaps

**Deborah Harry:** A tail might be nice. (Deborah Harry in MONDO 2000 magazine, 1991)

## Crazy Body Sculpting Desired

**Khannea Suntzu:** I want *crazy* body sculpting to begin as quickly as possible. I want these technologies tested by pioneers, thoroughly debugged, mass-produced, and safe, cheap, and far more user-friendly than any medical technology today.

Hell I want *massive market penetration*. Pardon my pun. I know what I am supposed to be, and that is a breathtakingly beautiful, tall, rather muscled and voluptuous, constantly-smiling woman in her mid-twenties; a fine blend of all the world's races with a subtle Mediterranean leitmotif, hauntingly soft blue eyes, and small horns and a nice hairless tail. Very robust throat, vagina, and ass, thank you very much. My current assets are already satisfying but I want them seriously younger, more elegantly feminine, healthier, and significantly upgraded.

## Imaginative Bio Body Sculpting

**Christopher Dewdney:** Most people's "ideals" would turn them into underachieving Nicole Kidmans and eight-foot Brad Pitts, identical cutouts. My previous, rather naïve, notion was that biotechnology would free us from the tyranny of "normalcy"—that we could become anything we wanted, morph ourselves into elongated, blue-skinned, orange-haired, sixteen-fingered geniuses, or perhaps flying ribbons of sensual bliss that performed acrobatic choreographies above the sunset.

# CALORIC RESTRICTION

*(See also Longevity/Immortality, Quantified Self)*

As characters in the movie *This Is Spinal Tap* once noted, "It's such a fine line between stupid and clever." For advocates of caloric restriction (CR), there may be a fine (or skinny) line between unhealthy anorexia and prolonged life.

## AN OLD IDEA DOES NOT NECESSARILY MAKE FOR AN OLD PERSON

CR is a longevity trope that has been around for a long time among transhumanists. The effects were first noted way back in 1934 when experimenters at Cornell found that rats lived twice their normal lifespan on a restricted diet. Roy Walford—an early life extension fanatic who also used to meditate while sitting on a block of ice—experimented with the method starting in the 1970s. He's dead.

## NOSFERATU-LOOKING BALDHEADED DUDES

By dint of observation at numerous transhumanist conferences, I can conjecture that, although the theory behind caloric restriction may be modestly popular, the practice is relatively rare. Most would probably prefer to wait for a magic pill that mimics CR's effects (see Nutraceuticals). However, there does seem to be a small number of extremely skinny bald-headed attendees who look like Nosferatu. I'm betting they're practitioners.

## What Roy Walford Learned

**Meredith Averill and Paul McGlothin:** Roy Walford, Ph.D. gerontologist and Father of CR for Humans, studied calorie restriction in mice at UCLA and was awarded many honors for his work. He saw that mice and monkeys, both mammals, shared greatly improved health and longer life when maintained on a CR diet.

Dr. Walford was a pioneer in the field of CR for human longevity. He was also the only physician in the Biosphere experiment, which turned into the first human CR experiment. Biosphere 2, which ran from September 1991 through September 1993, provided the first insights into the effects of CR on people. He learned that humans, when limiting their calories, also display the same effects on their blood-level markers that are characteristic of CR in the other two species of mammal.

## And This One Says It's Dangerous and Doesn't Work

**Athena Andreadis:** What is the basis for caloric restriction as a method of prolonging life? The answer is . . . not humans. The basis is that it appears (emphasis on the appears) that feeding several organisms, including mice and rhesus monkeys, near-starvation diets seems to roughly double their lifespan. Ergo, reasons your average hopeful transhumanist, the same could happen to me if only I had the discipline and time to do the same—plus the money, of course, for all the supplements and vitamins that such a regime absolutely requires, to say nothing of the expense of such boutique items as digital balances . . .

For women in particular, who are prone to both anorexia and osteoporosis, caloric restriction is dangerous—hovering as it does near keeling-over territory.

## And This One Says It's No Fun

**Hank Pellissier:** Daily, I ingest about 3,600 oily, spicy calories. Does this make me corpulent? No! I'm six foot two, 168 lbs, 9.9 percent body fat, cholesterol 145, BMI 21.6. My accelerated metabolism keeps me scrawny because I swim, lift weights, and run off an extra 4,000 calories a week. Plus I'm vegetarian, supplementing with vitamins, minerals, omega-3, and resveratrol (the longevity elixir).

Is this fine shape for a fifty-seven-year-old? The doc says yes. The wife responds. Everybody thinks I'm healthy, except . . . the CRONies (Caloric Restriction with Optimal Nutrition). They say I'm going to *die*. Before they do. Because I don't follow their diet. CRONies think they'll thrash my ass in the race away from the Grim Reaper.

What are these CRONies full of? Not much. CR's traditional goal is to reduce calories by 30 percent. US Government Recommended Dietary Allowances (RDA) is 2,000 calories per day; that means fully-dedicated CRONies nibble through life on 1,400 per diem—I out-eat 2.5 of them! Ha, ha. I enjoy that, but I'm annoyed by their death warnings. CR implies I should lose thirteen pounds to get my BMI down to 19.9. Grrr! I want to snag longevity by saying yes to endurance aerobics, super sex, tasty antioxidants, and resveratrol. I already said no to meat and ice cream, isn't that enough?

Not for the CRONies. Lisa Walford—daughter of Dr. Roy Walford, wife of Brian Delaney (president of Caloric Restriction Society) and co-author of *The Longevity Diet*—has a BMI of 15 and she described her breakfast as: four walnuts, six almonds, 10 peanuts. GASP! Where's my hash browns, ketchup, bagels, cream cheese, and huevos rancheros?

At PubMed I dig up an abstract from Boston noting that the suicide rate of men goes up when their BMI dips below 21. Yikes! CRON ambition is 20 or less.

CR doesn't offer enough gain for the pain. CRONies hope to extend their lives 15–25 percent from the day they begin, but most scientists view this as wildly optimistic. Dr. Phelan suggests longevity "might increase by 2 percent."

# CITIZEN SCIENTISTS

*(See also Biohacking, Grinders)*

The citizen scientist is someone who is doing scientific experiments outside the usual institutions and, in some cases, without the typical academic accreditation that makes someone an "official scientist." Citizen scientific work is generally open source (all information is freely shared). Biohacking, discussed above, is one aspect of citizen science.

# Early Citizen Scientists

**Joseph Jackson (interviewed by Alex Lightman):** A citizen scientist is anyone who uses the scientific method to investigate themselves or their environment to answer a particular question or satisfy their curiosity. Several exemplary historical citizen scientists come to mind. Thomas Jefferson is the archetype of the gentleman scholar. Benjamin Franklin invented bifocals when he got tired of switching between two pairs of glasses and, of course, famously flew a kite in a lightning storm to discover the principles of electricity. Edward Jenner discovered inoculation and performed the first vaccination against smallpox. Jenner's case is especially important as it highlights the power of user innovation. As a country doctor, Jenner observed that milkmaids who interacted with cattle infected with cowpox did not contract smallpox. He then transferred pus from a milkmaid to a young boy, completely protecting him from smallpox.

The medical establishment was reluctant to accept the findings of a "lowly" country physician, but eventually Jenner prevailed. Thomas Edison also partially fits the descriptor of citizen scientist but because he was, frankly, a bit of a bastard (see his feud with Tesla and other abusive monopolistic industrial practices), his example is not one we want to encourage under the new Open Science paradigm. Most importantly, in the 21st century, for the first time, the plummeting cost of technology enables anyone to be a citizen scientist, whereas the classic citizen scientists of the first Enlightenment were all wealthy men who had the time and resources to conduct experiments.

# CLONING

Cloning burst into public consciousness in the 1970s, with movies like *Sleeper* and *The Boys from Brazil*, and David Rorvik's "nonfiction" hoax book *In His Image: The Cloning of a Man*. Suddenly, a Brave New World of babies in beakers seemed terrifyingly close. Ever since, clones have been fodder for sensational headlines and a staple of science fiction and horror films.

However, clones are really nothing new. Identical twins are clones. Asexual reproduction produces genetically identical offspring: also clones. Parthenogenesis, done by many plants and some animals, produces offspring that are called full clones or half clones. When a worm is severed and the two halves regenerate into two worms, they're clones.

Even the artificial cloning of animals is older than the term "science fiction": Hans Driesch first cloned sea urchins from embryos in the 1880s, and Hans Spemann cloned a salamander in 1902.

## SOMATIC CELL NUCLEAR TRANSFER

These days, cloning is done through a process called somatic cell nuclear transfer (SCNT). An egg has its nucleus removed and replaced with the nucleus of an adult donor cell. This modified egg is stimulated with electricity or chemicals (often caffeine) to induce cell division.

There are four types of cloning.

### Reproductive Cloning

In reproductive cloning the modified cell is transferred to the uterus of a female, where it (hopefully) develops to term. Reproductive cloning reached a milestone with the world-famous sheep named Dolly, the first mammal to be cloned from an adult cell. Revealed in 1997, Dolly sparked a flurry of debate and numerous declarations and laws. So far, around two dozen animal species have been cloned, and not just as experiments: cows that are best at milk production are now often cloned. However, some species (e.g. monkeys) seem to be more resistant to SCNT.

Cloning of human embryos is not uncommon (see below), but despite periodic claims of cloned babies, none have been confirmed. Reproductive cloning would be the method used for resurrecting extinct species, à la *Jurassic Park*.

### Therapeutic Cloning

Also called embryo cloning, this differs from reproductive cloning in that the cloned cells remain in the lab, and so aren't allowed to develop into a complete organism. It's used to grow stem cells used in research and for creating replacement tissues for burn and accident victims. Unfortunately, extracting the stem cells destroys the embryo, raising ethical concerns for some.

### Replacement Cloning

This is a theoretical combination of the first two kinds: a form of regenerative medicine using a cloned copy to replace all or part of a body, by using either a brain transplant or by harvesting the body parts of the clone. As far as is known, this has never been accomplished.

### Gene Cloning

Also called DNA cloning or molecular cloning, this is done by moving a DNA fragment from one organism to a foreign host cell. It's used by researchers to generate multiple copies of the same gene.

## GENETIC SAVINGS AND CLONE

Once upon a time, a fellow named Lou Hawthorne was hired by an unidentified wealthy woman to clone her recently deceased dog, Missy. The cloning was successful, and it generated a small wave of controversy and hype, leading to the formation of Genetic Savings and Clone in Sausalito, California.

In 2000, they succeeded in cloning a cat, who they named "CC" (presumably for Cloned Cat). In 2004, GS&C cloned another cat, Little Nicky, its first commercial product. Unfortunately for GS&C, but fortunately for animals waiting in shelters, business in cloned pets was anything but brisk, and the company closed its beakers in 2006, leaving pet lovers who had sent them their pet's DNA in the lurch. (Though GS&C promised to forward the DNA to another similar facility, if such a thing existed.)

## ETHICAL AND LEGAL ISSUES

Critics see many types of cloning as dehumanizing. In a nonbinding declaration, the United Nations has called for banning all forms of human cloning and all "genetic engineering techniques that may be contrary to human dignity." The Catholic church and Sunni Muslims oppose it, though Iran (which is Shiite Muslim) is using cloning as part of a drive to become a biotech leader. California banned human reproductive cloning in 1997, and a dozen other states and numerous countries have also done so. The Humane Society of the United States and the American Anti-Vivisection Society have denounced commercial pet cloning as cruel, but it remains largely unregulated.

Transhumanist uses of cloning may or may not hit ethical or legal roadblocks. Replacing damaged or worn tissues or body parts with new cloned replacements should be fine, especially if a way could be developed that did not destroy embryos. However, growing an entire replacement clone body would be problematic, ethically and practically. After all, your clone is exactly as human as you are. Cloning yourself and then harvesting the body for your own use is far too supervillainish for most of us, and the law would not approve. Besides, transferring your elderly brain into the youthful body of your clone won't make your brain any younger. Also, since it would take around eighteen years to grow an adult clone, and to get a properly developed one, you couldn't just leave it in a giant beaker the whole time, so you'd have to raise it like any other child. Given the investment of that much time and effort, even a supervillain might get emotionally attached, and call off the body harvesting plan.

# COGNITIVE ENHANCEMENT

*(See also Implants, Nutraceuticals, Neurotechnology, Psychedelic Transhumanism)*

While making smarter-than-human machines is a big focus for many on the transhumanist edge, most have not given up on upgrading our own cognitive capacities. The idea that there are substances—drugs or nutrients—that increase human intelligence goes back at least to the 1960s, when psychopharmacological expert Dr. Nathan Kline predicted that big IQ drugs were right around the corner. By the 1970s, an odd couple who looked like heavy metal rock stars, Durk Pearson and Sandy Shaw, were pitching intelligence increase through nutrient formulas as well as through drugs as regulars on TV talk shows. Timothy Leary (see entry) also added intelligence increase to his bag of magic tricks.

During the '90s, it seemed like the age of smart drugs and nutrients was upon us, but the energy dissipated as it was noted that the newly available (by mail order) substances like vasopressin and piracetam had stimulant effects, but didn't make us all instant geniuses. Caffeine and nicotine, which work by tweaking our neurons—in the case of coffee, by inhibiting our inhibitory neurotransmitters—are still the most commonly used cognitive enhancers.

Today, a new breed of neuroenhancers is starting to appear with increasing frequency on university campuses around the world. *Nature* reports that students are striking deals to buy and sell prescription drugs such as Adderall, Ritalin, and modafinil—not to get high, but to increase their capacity for learning and get higher grades.

Nonprescription sale and use of these drugs are crimes in the United States, punishable by prison.

**Surfdaddy Orca:** Modafinil—a banned stimulant in competitive sports—enhances academic productivity and significantly reduces the need for sleep to a couple of hours per night while improving working memory. A University of York website describes three students—Charles, Nick, and David—who each took a 200 mg tablet of modafinil. According to Charles, "After an hour, none of us felt any different. But then I started to feel markedly more alert. I couldn't be sure it wasn't a placebo, but then Nick became uncannily good at computer games, beating his friends three times in a row at Pro Evo. It was no coincidence."

Modafinil has proven so popular in the academic pressure cookers of Oxford and Cambridge that close to one in ten students have admitted taking prescription medication such as modafinil without the prescription. The academic uses range from increased alertness during exams to stimulating thought processes when writing essays or take-home exams.

In short, many of today's students would rather drop modafinil than LSD, preferring a competitive edge to an expanded mind. But do drugs like Adderall, Ritalin, and modafinil really enhance intelligence, increase focus, and boost creativity? Bruce Katz comments, "As far as increasing intelligence, this is a . . . difficult matter. For example, simply increasing the brain's learning rate may speed up the acquisition of new concepts, but will also increase the rate of catastrophic forgetting of older concepts. Intelligence and wisdom is not just about knowledge acquisition, but in applying this knowledge in the right contexts."

## Creativity Enhancement?

**Bruce Katz (interviewed by Surfdaddy Orca):** There are two primary types of cognitive enhancement—enhancement of intelligence and enhancement of creative faculties. Even though creativity is often considered a quasi-mystical process, it may surprise some that we are actually closer to enhancing this aspect of cognition than pure intelligence.

The reason is that intelligence is an unwieldy collection of processes, and creativity is more akin to a state, so it may very well be possible to produce higher levels of creative insight *for a fixed level of intelligence* before we are able to make people smarter in general.

There appear to be three main neurophysiological ingredients that influence the creative process. These are 1) relatively low levels of cortical arousal; 2) a relatively flat associative gradient; 3) a judicious amount of noise in the cognitive system.

All three ingredients conspire to encourage the conditions whereby cognition runs outside of its normal attractors, and produces new and potentially valuable insights.

What are the implications for artificially enhancing insight? First, any technique that quiets the mind is likely to have beneficial effects. These include traditional meditative techniques, but possibly also more brute-force technologies such as transcranial magnetic stimulation (TMS). Low frequency pulses (below 1Hz) enable inhibitory processes, and TMS applied in this manner to the frontal cortices could produce the desired result.

Second, the inhibition of the more literal and less associative left hemisphere through similar means could also produce good results. In fact, EEG studies of people solving computer science problems with insight have shown an increase in gamma activity (possibly indicative of conceptual binding activity) in the right but not the left hemisphere just prior to solution.

Finally, the application of noise to the brain, either non-invasively, through TMS, or eventually through direct stimulation, may encourage it to be more "playful" and to escape its normal ruts.

# COGNITIVE SCIENCE

*(See also Artificial Intelligence, Distributed Cognition)*

## THINKING ABOUT THINKING

Cognitive science is the study of how information is represented and computed, in human and animal minds and in computers. It's a large, multidisciplinary field with many levels, and includes aspects of neuroscience, psychology, linguistics, artificial intelligence, philosophy, and anthropology.

## REPRESENTATIONS AND COMPUTATIONS

Cognitive science views the mind in terms of mental representations and the computations that process them. As you might expect, the nature of these representations and computations is endlessly disputed, but the current approaches can be summarized as formal logic, rules, concepts, analogies, images, and connectionism. All those paradigms are valuable, but connectionism is the newest and perhaps the most interesting. Connectionism does not model the mind in the standard way, as a sort of computer processing a symbolic language.

## SIMPLE + CONNECTED = EMERGENT

We know that the vast powers of the human mind arise from relatively simple and uniform units (neurons) and their interconnections (synapses), all operating in parallel. Neural networks, now the most common connectionist model, are designed to mimic this. Thought processes are modeled by activations spreading across units via the connections, with the pattern determined by the weights (or connection strengths) between units. In other words, information is not processed and stored as strings of symbols like a book or a computer program, but non-symbolically, in the weights of connections.

Neural network models offer many advantages. They're flexible, good at dealing with fuzzy and difficult problems, and at handling noise and damage. They're used for things like pattern recognition, computer vision, image compression, and identifying new molecules for use in drugs.

## CRITIQUES

Some critics believe the representation and computation model is flawed. They feel cognitive science neglects things like emotions, consciousness, the physical environment, the human body, and society, or that the mind isn't really a computational system, but a dynamical one. However, others believe these aspects can be integrated into the representation and computation model.

## BETTER BRAINS, PHYSICAL AND VIRTUAL

Transhumanists are interested for several reasons. Understanding cognition is key to preserving and enhancing brain function: it helps to know how things work if you want to improve your IQ or memory. (Not to mention if you want to upload your mind to a computer.) It's also key to creating any form of artificial intelligence, so cognitive science would be crucial to reaching The Singularity.

### Cognitive Science Is Not for the Thin Skinned

**Joe Quirk:** If you want to believe your own bullshit, stay away from cognitive science . . . Hey numbnuts, cognitive science demonstrates that you're not bright enough to realize what a clusterfuck your life is, because you're wired to tell yourself a coherent story after the fact. Microsecond by microsecond, your neocortex spins a story that says: "I meant to do that." Your conscious mind thinks it's Sherlock Holmes, but really it's Maxwell Smart, tripping through life and weaving coherent excuses to maintain the illusion of control.

# CONSCIOUSNESS

If we're going to build extremely intelligent machines, they need to be conscious. Or do they? Even today, after decades of extensive research in cognitive science, nobody can say precisely how consciousness arises. Many scientists will tell you we can't even say what it is, exactly. The best definition may be that consciousness is being aware of being aware, or being aware of having an experience. In other words, subjectivity. A *New Scientist* article from 2009, covering the work of French neuroscientist Raphaël Gaillard, states: "Signals from electrodes seem to show that consciousness arises from the coordinated activity of the entire brain. The signals also take us closer to finding an objective 'consciousness signature . . .'"

If we want unimaginably brilliant machines to serve us well, we may not want them to be fully conscious, to be aware of having experiences. But then the question becomes whether certain types of useful intelligence aren't, in fact, built upon the desires aroused by being aware of having an experience.

## The Great Consciousness Swindle

**James Kent:** The great Consciousness swindle is the assumption that "Consciousness," with a capital C, is so complex and mysterious that stupid blind neuroscientists can never explain it all with their crude, classical, materialistic descriptions. This, of course, is a complete intellectual fallacy. Scientists who study the brain understand that "consciousness," with a lowercase c, is not a "thing" with a "location," but is instead the abstract process of being self-aware, or a relative measurement of general self-awareness.

When you talk about consciousness with a lowercase c, then it becomes easy to see that consciousness is not mysterious at all. It is a description of our everyday waking lives. For humans, consciousness comes online when we wake up and goes through peaks and valleys throughout the day. Consciousness gets hungry, tired, bored, excited, aroused, irritated, distracted, and so on, until we go back to sleep and consciousness disappears and we become "unconscious." Then consciousness comes back online in a very limited "secure test environment" for a few seconds at a time while we dream, then it disappears again. And when we wake up the cycle resets and consciousness starts a new day.

## Kurzweil Can't Get No Satisfaction from Theories of Consciousness

**Ray Kurzweil (interviewed by Surfdaddy Orca and R.U. Sirius):** I get very excited about discussions about the true nature of consciousness, because I've been thinking about this issue for fifty years. And it's a very difficult subject. When some article purports to present the neurological basis of consciousness . . . I read it. And the articles usually start out, "Well, we think that consciousness is caused by . . ." You know, fill in the blank. And then it goes on with a big extensive examination of that phenomenon. And at the end of the article, I inevitably find myself thinking . . . where is the link to consciousness? Where is any justification for believing that this phenomenon should cause consciousness?

My thesis is that there's really no way to measure consciousness. There's no "Consciousness Detector" that you could imagine where the green light comes on and you can go, "OK, this one's conscious!" or, "No, this one isn't conscious."

Even among humans, there's no clear consensus as to who's conscious and who is not. We're discovering now that people who are considered minimally conscious, or even in a vegetative state, actually have quite a bit going on in their neocortex and we've been able to communicate with some of them using either realtime brain scanning or other methods.

Consciousness is the "hard problem" in mind science: explaining how the astonishing private world of consciousness emerges from neuronal activity.

# COSMISM

The term "cosmism" was originated by Konstantin Tsiolkovsky and other Russian futurists in the late 1800s, and then borrowed by Ben Goertzel and Giulio Prisco in 2010 to denote a futurist philosophy more tailored for the modern era.

Today's cosmism posits a positive, far-reaching, blatantly transhumanist attitude toward science, technology, life, the universe, and everything. As summarized in Goertzel's 2010 book *A Cosmist Manifesto*, contemporary cosmism is less an analytical philosophical theory, and more an everyday sort of philosophy, focused on enthusiastically and thoroughly exploring, understanding, and enjoying the cosmos, and being open to all the possible forms life and mind may take as the future unfolds.

Cosmism advocates:

- Pursuing joy, growth, and freedom for oneself and all beings.

- Ongoingly, actively seeking to better understand the universe in its multiple aspects, from a variety of perspectives.

- Taking nothing as axiomatic and accepting all ideas, beliefs, and habits as open to revision based on thought, dialogue, and experience.

Near the beginning of *A Cosmist Manifesto* is a list of principles, formulated by Goertzel and Prisco, which serve as the heart of cosmist thinking. We're sharing some bits from them with you:

- Humans will merge with technology, to a rapidly increasing extent. This is a new phase of the evolution of our species, just picking up speed about now. The divide between natural and artificial will blur, then disappear. Some of us will continue to be humans, but with a radically expanded and always growing range of available options, and radically increased diversity and complexity. Others will grow into new forms of intelligence far beyond the human domain.

- We will develop sentient AI and mind uploading technology. Mind uploading technology will permit an indefinite lifespan to those who choose to leave biology behind and upload. Some uploaded humans will choose to merge with each other and with AIs. This will require reformulations of current notions of self, but we will be able to cope.

- We will spread to the stars and roam the universe. We will meet and merge with other species out there. We may roam to other dimensions of existence as well, beyond the ones of which we're currently aware.

- Spacetime engineering and future magic will permit achieving, by scientific means, most of the promises of religions—and many amazing things that no human religion ever dreamed. Eventually we will be able to resurrect the dead by "copying them to the future."

- Intelligent life will become the main factor in the evolution of the cosmos, and steer it toward an intended path.

Cosmism is a sort of philosophically laid-back version of transhumanism. In a culture that tends to be argumentative and filled with people who like to insist that their views are correct, cosmism doesn't care if you're viewing the universe as information or quantum information or hypercomputation or God stuff or whatever. Nor does it ask anyone to commit to AGI or mind uploading or brain-computer interfaces or fusion-powered toasters as the best way forward. Rather, it seeks to infuse the human universe with an attitude of joy, growth, choice, and open-mindedness. Cosmism believes that science in its current form, just like religion and philosophy in their current forms, may turn out to be overly limited for the task of understanding life, mind, society, and reality—but it teaches that, if so, by actively engaging with the world and studying and engineering things, and by reflecting on ourselves carefully and intelligently, we will likely be able to discover the next stage in the evolution of collective thinking.

# CRITICISMS OF TRANSHUMANISM

*(See also Designer Babies)*

Empowering individuals and transcending what were long considered human limits: these goals are exciting to some, but they're disturbing and frightening to others. Let's put major objections to transhumanism into one or more of four categories: that it's unfeasible, directly dangerous, indirectly dangerous, or immoral.

## "IT WON'T WORK!"

It's easy to be skeptical about technological predictions. After all, nobody commutes by personal helicopter or nuclear-powered automobile, or has a kitchen robot cooking dinner. Nuclear power never made electricity "too cheap to meter," and controlled fusion has been twenty years away for about sixty years now . . . and still is.

Some say that mind uploading is impossible, pointing to the assumptions implicit in the concept: the idea that "you" (your self, mind, or soul) is something distinct from your body. This philosophical dualism is seen as inconsistent with the materialism transhumanists otherwise profess.

But let's assume it's perfected. You're on your deathbed with a terminal illness, your mind is copied to a computer, and your body dies. Are "you" now in the computer? Or are you dead, and what's in the computer is just a copy of "you"? That's nice for the copy that's happy to be there (one hopes), but it's not much consolation for the old you that was in your body. So mind uploading can be seen as an ontological fail.

Core aspects of The Singularity, and Kurzweilian techno-optimism in general, have many critics. Paul Allen points out that as we gain deeper knowledge of natural systems like human intelligence, our theories become increasingly complex. Moore's Law notwithstanding, the Law of Accelerating Returns thus runs up against this "complexity brake." Since the concept of artificial intelligence is central to The Singularity, the complexity brake may delay it, if not make it forever unobtainable.

## "IT'S DANGEROUS AND COULD KILL US ALL!"

All tools can be used for good or evil, and transhumanist technology has a number of what might be called direct dangers. 3D printers can print unregistered guns for the wrong people. Nanotech molecular assemblers could become tabletop drug labs. Mind enhancement could be used by criminals to enhance their criminal abilities. Biotechnology could be used to create terrorist bioweapons.

Some fear that GMOs could disrupt the world's food supply. Self-replicating nanobots could get out of control and consume everything on Earth: the "grey goo" scenario. A powerful, post-Singularity artificial intelligence could decide to exterminate humanity—the Skynet scenario from the Terminator movies.

## "IT'S SOCIALLY DESTRUCTIVE!"

The indirect dangers of transhumanism are at least as numerous. Sometimes the fears are about society or the state being damaged by out-of-control technology in the hands of individuals. At other times, the fears are that society or the state will gain too much power over individuals.

Although the old-fashioned coercive eugenics of the early 20th century progressive and Nazi varieties are long gone, their ghosts still haunt any discussion of transhumanism. Critics warn of socially divisive effects, such as a "genetic divide" of classes based on genetic modifications, as portrayed in the dystopian movie *Gattaca* or the novel *Brave New World*. Francis Fukuyama has called transhumanism "the world's most dangerous idea," seeing even voluntary genetic improvements as an anti-egalitarian threat to the ideals of liberal democracy.

After all, the benefits of transhumanism aren't going to happen everywhere and to everyone all at once. Some people will be the first to benefit from brain enhancements and life extension, and those people are likely to be rich and have the right social or political connections. Social cohesiveness could suffer. Imagine mentally enhanced, immortal rich people and politicians, whose enhanced offspring always get the best grades, the best jobs, and win every Oscar, Emmy, and MVP award. Nobody wants a transhuman overclass, except the people who imagine themselves in it.

Or, perhaps even worse, the blurring of the meaning of "human" could lead to the creation of an underclass of semi-humans. Think *The Island of Doctor Moreau*, *Blade Runner* replicants, or Cordwainer Smith's Underpeople.

## "IT'S DEHUMANIZING AND PLAYING GOD!"

More philosophical objections come from critics who agree on little else, but see transhumanism as essentially immoral and anti-human.

The traditionalist, conservative attack can be summed up in a sentence from a 2002 Vatican statement: "Changing the genetic identity of man as a human person through the production of an infrahuman being is radically immoral." For many Christians (and members of other Abrahamic faiths), transcending human limits is a dangerously hubristic idea, because it gives humans powers that should be reserved for God. Transhumanism is seen as a form of "scientism"—a dogma that empirical science is the most authoritative worldview, and that only measurable knowledge is valuable. Thus, transhumanism is a type of idolatry and more "false good news," a utopian movement trying to immanentize the eschaton (create heaven on earth). Conquering death and creating utopia cannot happen through technological means, they say, but only through God. Furthermore, God doesn't appreciate the attempts at competition.

There are related attacks from the secular left: transhumanism is seen as "atomized individualism," power fantasies, and as an expression of contempt for the flesh. Feminists may see extensions of unhealthy cultural obsessions with youth and beauty. Some consider biological enhancements as trivializing human identity, or see an "ableist bias" in even thinking in terms of "improvements" or overcoming mental or physical "limitations."

## POINTS TAKEN, BUT . . .

For some, possible dangers will always trump possible benefits, and alleviating existing poverty or inequality will always take precedence over an advancement that will benefit only the relatively wealthy (at least at first).

Of course, taken to its logical extreme, those views would stop most technological advancements, because they all have some dangers and negative side effects. The rich or the lucky few always benefit first, so types of inequality will temporarily increase.

Besides, being first isn't always best. Think of those cutting-edge LED watches that cost hundreds or thousands of dollars in the early '70s. That money helped finance the cheaper and more advanced digital technology that came later, and all those rich people got was a couple of years of looking fashionable, and they had to press a button to see the time.

As we saw with thalidomide and metal-on-metal hip implants, being an early adopter can be risky. So perhaps we should not worry about the rich financing transhumanist biotech. They'll be the first guinea pigs, and you can get it when it's cheaper, better, and safer.

If humanity had always viewed technology through the lenses of the precautionary principle and equal access, we'd still be arguing whether to allow this invention called "fire."

# CRYONICS

The embrace of cryonic preservation may be the thing that most separates "the men from the boys" (pardon my genderism) in terms of how seriously you'll be taken by hardcore transhumanists. Most major transhumanist figures are signed up.

Cryonics, of course, is the use of low temperature to preserve bodies after death (or as former cryonics enthusiast Timothy Leary preferred to call it, "reversible metabolic coma") for their eventual resuscitation. It comes in two forms—whole body or head. The idea is that the personality is preserved in brain structure and that if you can freeze it fast enough, future science may be able to bring back the same person with his or her memories preserved.

## CORPSICLES

Cryonics has what transhumanists call a high "yuck factor." People tend to be repulsed, even before examining it. Even among technophiles, the snarky label "corpsicles" is often used.

## ALCOR AND CONTROVERSY

There are a number of cryonic outfits around the world, but the one most closely associated with the transhumanist movement is Alcor Life Extension Foundation in Scottsdale, Arizona. Alcor has 117 patients currently on ice.

Unsurprisingly, given the nature of its work, the company has been wrapped up in a few controversies. In 1994, the necessity to get a patient into preservation ASAP led to charges of murder, when barbiturates were given to Dora Kent in what the coroner suspected was an act of euthanasia prior to her natural death. However, no charges were ever filed.

In 2003, there were allegations that the frozen head of baseball great Ted Williams, Alcor's most famous deceased client, had been abused. It was, in all likelihood, a misunderstanding of cryonic preservation procedures. Cryonicists may be a little odd, but they don't fuck around, which, actually, is why Timothy Leary—a lapsed customer of Alcor and then CryoCare (he chose to be cremated)—decided he didn't like them.

## STILL NOT READY FOR PRIME TIME SCIENCE

Most medical scientists dismiss cryonics as a fantasy, although cryogenics are starting to be used in medicine, to preserve body parts among other things. The question that haunts cryonics is not so much whether a body might be able to be brought back to life, but whether the brain's memories will be preserved or will succumb to freezer burn. Literally.

## The Simple Premise

**Sandy Sandfort:** The premise of cryonics is simple: do nothing, and when you die, you die forever; get cryopreserved, and when you die, maybe you get a second chance. But what does "when you die" mean, and why does it matter?

In the past, if your heart stopped beating and your lungs stopped breathing—end of story, you were legally dead. Then resuscitation technology got better and "dead" people starting rebooting. Ta-da!

So the goal posts were moved and death became popularly defined as "brain death." Surely, you are dead if you have no measurable brain activity. Nope. People have already been brought back from that. Now, cryonic proponents say you aren't dead . . . really dead . . . until you have suffered information-theoretical death.

Whatever death is taken to mean, one thing is obvious. The "fresher" you are when cryopreservation begins, the better your chances of returning. The ugly truth is that the moment you die—by any definition of death—you start to spoil. Every second that ticks by increases the amount of damage your body and brain undergo. Finally, at some point, you will be gone, no matter how advanced the technology.

## Progress

**Ben Goertzel:** Cryonics—the practice goes back at least to 1967, when the Cryonics Society of California cryopreserved James Bedford. But the technology has progressed tremendously since then, with organizations like Alcor and the Cryonics Institute making use of advanced techniques for preserving cryonics patients with less and less damage, increasing the odds of eventual successful resuscitation. Most exciting has been the development of vitrification, which allows the preservation of the body in a special glassy state, avoiding the damage ensuing from the traditional freezing process. And millions of dollars are being spent to keep the science and practice of cryopreservation moving forward.

# CYBORGS

*(See also Exoskeltons, Grinders, Implants, Robotics)*

A cyborg, or cybernetic organism, is a living thing that is a combination of organic biology and artificial technology. Many of our pets have microchip implants that help us locate them. They're cyborgs. My mother is a cyborg. She's had a pacemaker for twenty-five years. Every once in a while, it has to be replaced. Today we have all kinds of artificial parts. Anyone with an artificial part or even a frivolous implant is a cyborg.

## CYBORG FEMINISM

The notion comes from the influential work "The Cyborg Manifesto" by Donna Haraway, in which she embraces extreme technology as a means of liberating women from conventional roles. In a more complex vein, she argues, in essence, that the coming of the cyborgian woman could be revolutionary from a left radical perspective because it confuses conventional narratives of identity. Her most famous quote (a challenge to another mode of feminism) is "I'd rather be a cyborg than a goddess."

## Have a Turbine Heart

**Valkyrie Ice:** We as transhumans will have to overcome a human bias to actually become "Trans" humans.

Which bias is that, you ask? The idea that the human body as it currently is constructed is either "perfect" or that any "enhancements" must mimic how the body currently functions. I can remember the projections once made about the Jarvik heart, including the "commercial" that made it into *Robocop* that satirically predicted the "Jarvik Sports Heart" for the athletic heart patient. Yet here we are in the future predicted to have completely replaced transplants with engineered replacements, and the artificial heart that "beats" is still a fantasy. Why?

You might as well ask why we don't yet have airplanes with flapping wings. Then ask yourself why nature never evolved birds capable of flying faster than sound. The answer is that nature doesn't always come up with the "best solution"—just one that works. Just like Leonardo's flapping machines never flew, a beating heart has not merely proven exceptionally difficult to reproduce, but has proven to be needlessly complex in comparison to the likely future solution, a heart that has no beat, no pulse, and which pumps blood in a continuous flow, via turbine based "jets."

## Eyeborg

**Kristi Scott:** According to Rob Spence, who had a prosthetic camera inserted in the socket of a blind eye, the idea was quite natural to him. He notes that anyone who's seen *The Dark Crystal* knows the concept has been out there. He's been thinking about it for years. After it was all said and done, Rob admitted, "I'm happy that I lost the eye . . . I get the chance to stand out . . . as an augmented filmmaker."

Rob claims to have gotten lots of support from people with one eye . . . even old ladies. Apparently when these ladies call Rob they don't want something that looks normal. The general feedback he's getting from the older demographic is: "I can stick this in and be the envy of people and be more than I was."

## Cyborg vs. Goddess

**Kyle Munkittrick:** Donna Haraway's "Feminism in the Late Twentieth Century" is the *locus classicus* of cyberfeminism. Published as an essay in 1985 and then redrafted as a chapter in Haraway's *Simians, Cyborgs, and Women: The Reinvention of Nature* in 1991, the manifesto has aged particularly well, remaining relevant within feminism and cultural studies, and it is often quoted in transhumanist works. The manifesto was written as a rebuttal of eco-feminism, a philosophy that views technology as inherently patriarchal and advocates communism and deep ecology as a counterpoint to what they see as the Western capitalist patriarchy. Drawing partially upon Michel Foucault (whom she also mocks), Haraway argues instead that the very forms of power used by hegemonic forces can be used for resistance and liberation . . .

# DARPA

*(See also Brain-Building Projects, Exoskeletons, Warbots)*

When it comes to way mad science, science fiction can never top the US government's Defense Advanced Research Projects Agency (DARPA). For starters, its progenitor, ARPA, started the Internet (originally the ARPANET). In more recent times, DARPA has also been a progenitor of cloud computing, virtual reality, Apple's Siri speech recognition, self-driving cars, and onion routing—the thing that makes Internet anonymity possible. (DARPA, the NSA would like to have a word with you.)

## GONZO SCIENCE WITH GOVERNMENT SUPPORT

But those are just some of the more prosaic activities that DARPA is engaged in. Some of their projects point toward The Singularity (SyNAPSE program to build an artificial brain "that scales to biological levels," cryonic hyperlongevity) and other dream goals of transhumanism. Or as Surfdaddy Orca once put it in the title of an article, "DARPA Takes on Suspended Animation: Zombie Pigs, Squirrels, and Hypersleep." DARPA lends itself to hysterical tabloidesque fantasies.

Oh, to be DARPA and get to throw umpty-million dollars at far-out ideas (but—on the other hand—maybe not have it in the hands of the military).

### And in the Beginning

**Michael Belfiore (interviewed by R.U. Sirius):** DARPA started life as ARPA (without the "D" for defense) in response to the Soviet Union's 1957 launch of Sputnik. It was conceived as America's first space agency. But when NASA came online just a few months later, ARPA nearly didn't survive. It survived by branching out into other areas besides space, including information technology. And it survived by becoming a sort of dumping ground for projects that the armed services either didn't want or didn't know what to do with. That was how ARPA got its first computer, a surplus 250-ton, room-filling machine made to analyze radar data for the Air Force, along with the support staff needed to operate it. From there ARPA began its experiments in time sharing, interactive computing, and the ARPANET that was the genesis for today's Internet.

## Far Out!

**Michael Belfiore (interviewed by R.U. Sirius):** Two of the most "out there" projects I encountered were projects for modifying insects to act as spy bots, and one for programming matter to dynamically change shape, color, or just about any other property.

Programmable matter technology is in the very early stages, but the potential here is to create objects on the fly in response to user input or environmental conditions. Think the evil terminator in the second *Terminator* movie.

**Surfdaddy Orca:** DARPA is committing $9.9 million to the Texas A&M Institute for Preclinical Studies (TIPS) to research how hydrogen sulfide can block the body's ability to use oxygen and induce a state of suspended animation. Because they [pigs] have a cardiovascular system similar to humans, TIPS researchers Theresa Fossum and Matthew Miller think they can accurately predict human results from trials with anesthetized pigs. Using swine, the researchers are testing various compounds—some containing hydrogen sulfide—to find one that can safely keep the hemorrhaging animals "as close to death as possible."

**R.U. Sirius:** A DARPA gambit, as revealed by *Wired*'s Danger Room, is to "Assemble the latest biotech knowledge to come up with living, breathing creatures that are genetically engineered to 'produce the intended biological effect.' DARPA wants the organisms to be fortified with molecules that bolster cell resistance to death, so that the lab-monsters can 'ultimately be programmed to live indefinitely.'" In other words, immortal artificial biology.

# DESIGNER BABIES

*(See also Abolitionism, Criticisms of Transhumanism, Genomics)*

It's conjectured that not very far into the future, parents (or perhaps governments or corporations) will be able to precisely preselect traits for their offspring prior to birth via germline engineering: "designer babies."

A form of "designer baby making" that's already in use (pre-implantation genetic diagnosis) can be done only in cases that involve in-vitro fertilization. It is mainly used to screen for genetic diseases. Also, using in-vitro fertilization, it has been possible, for nearly ten years, for parents to select their baby's gender.

## UH-OH, EUGENICS!

Of all the tech developments in the transhuman armament, nothing freaks people out more than "designer babies." First of all, it's eugenics, and eugenics is reflexively associated with Nazism. The Nazis weren't the only ones obsessed with eugenics during the first half of the 20th century—it was going around. But, with their focus on racial purity and supremacy and the resultant horrors, their primitive version of designer babymaking insured that any thought of scientific experimentation leading toward "superhumanity" would be presumed to be venal. The other fear is that "designer babies" will only be available to the wealthy. The tremendous class divide that currently haunts the world would become a species divide—with rich people having smarter, healthier, more attractive, even posthuman children and the rest of us left behind. This is a valid fear, but one that might be met with social reform, rather than with knee-jerk anti-scientific bias.

# The Great Designer Baby Controversy of '09

**Michael Anissimov:** In early February 2009, the Fertility Institutes created enormous controversy by announcing that they planned to offer Pre-implantation Genetic Diagnosis (PGD) services allowing for the selection of eye and hair color for children. Dr. Jeff Steinberg was quoted by the BBC as saying, "I would not say this is a dangerous road. It's an uncharted road." As a scientist experienced in PGD/IVF techniques, Steinberg was aware that the technology to select physical traits in humans had been available for years, but no one would touch it. "It's time for everyone to pull their heads out of the sand," Steinberg said.

Transhumanists and other fans of procreative freedom were excited by the news.

The backlash was widespread. Quoted in the *New York Daily News* on February 23, 2009, the Pope himself condemned the "obsessive search for the perfect child." The pontiff complained, "A new mentality is creeping in that tends to justify a different consideration of life and personal dignity." The Roman Catholic Church objects to all applications of PGD because they invariably involve the destruction of blastocysts.

Then, all of a sudden, on March 2, Steinberg capitulated to widespread criticism. A press release on the Fertility Institutes website read, "In response to feedback received related to our plans to introduce preimplantation genetic prediction of eye pigmentation, an internal, self regulatory decision has been made to proceed no further with this project." *Gattaca* was averted.

# DISTRIBUTED COGNITION

*(See also Artificial Intelligence, Cognitive Science)*

## THE EXTENDED MIND

The theory of distributed cognition (also known as DC, DCog, or socially distributed cognition) was first developed in the 1980s by Edwin Hutchins, a cognitive psychologist and anthropologist who studied how navigation was coordinated on US Navy ships. He realized that cognition is best understood as something that happens not just in an individual brain, but something that encompasses knowledge placed in objects and individuals across the physical and social environment.

All this time, you thought thinking was a solitary activity, but it's really a collaborative ecosystem of specialized units.

Seen this way, the transhumanist vision of the extended mind or hybrid self has existed since humans evolved. It took huge leaps with inventions like the alphabet, the printing press, the telescope, and the telephone. Sensors and methods of communication and data storage aren't simply convenient tools: everything from collaborative tagging on websites, to smartphones, quadcopter surveillance, Wikipedia, World of Warcraft, Google, and the NSA, are now all changing the ways we think. Developments such as augmented reality, artificial intelligence, and neural interfaces will change us even more.

**James Kent:** Any cognitive system inherently relies on distribution, or the off-loading and coordination of specialized tasks to perform a complex function; whether it be a brain, a hive of ants, a classroom, or a cockpit full of pilots. The new school of Distributed Cognition (DC) says the rules for analyzing all of these distributed systems are inherently the same.

Since the elements of distributed cognition include individuals, groups, and their environment, defining the boundaries of DC seems like an arbitrary task. Technically, every form of cognition is distributed across some systemic framework, and the primary source of information and feedback control for all known cognitive systems is reality, a data set that contains the entire universe in motion. Where precisely do the boundaries of DC start and end? At the cellular level? The species level? The tribal level? The family level? The galactic level?

To break down the fundamentals of DC, the current thinking is that any set of cognitive units that work in groups (neurons, ants, people) will naturally fall into states of specialized cooperation that maximize the energy efficiency and output of the entire system. These states of specialized co-operation can be studied in terms of input and output to see what kind of results and errors emerge at the group level that would not exist if all parts of a system performed the same tasks in serial or parallel. The performance variables between emergent (dynamic) and aggregate (unchanging) states in distributed systems can be applied equally to paths worn into the lawns connecting the buildings of university campuses or the corporate-mandated behaviors of a team of baristas filling orders at your local coffee shop. The explicit rules that govern group behavior are not always most efficient for the individual, and thus improvised shortcuts are taken and new group behaviors emerge from the bottom up. These energy saving shortcuts are not evidence of laziness or bending the rules, they are evidence of distributed cognition at work. See, it's easy.

# EVOLUTIONARY PSYCHOLOGY

Fundamentally, evolutionary psychology proposes that our internal mechanisms and our behaviors are largely a product of biology—adaptations brought about through natural selection. The field has been met with a lot of angry resistance, primarily from those on the political left and from feminists and others who resist the idea that different tendencies among genders are heritable.

It seems to us that no educated person would deny that the behaviors and (in some cases) gender differentiations in other species are a product of biology, so only those with an ideological or religious bias can assume that humans have somehow completely escaped this fate.

On the other hand, evolutionary psych theorists get into trouble when they think they have us—and all our choices—totally sussed, based on genetics. Evolutionary psychology is a work in progress, and most advocates will tell you that the influences on our internal states and behaviors are about evenly split—half nature and half nurture. Evolutionary psych needs to be a bit cautious with declarations about genetic predispositions in humans because our ideas about how biology influences psychology and behavior are incomplete.

Transhumanists tend to embrace evolutionary psych, but also see themselves as hacking (trying to overcome) the diktats of Darwinian biology.

**David Pearce:** Evolutionary psychology explains how our moral intuitions and the rationalizations they spawn have been shaped by millennia of natural selection to maximize the inclusive fitness of our genes and not to track the welfare of other sentient beings impartially conceived. Many human cultures have found nothing intuitively wrong with aggressive warfare, slavery, wife beating, infanticide, or female genital mutilation. Ultimately, folk morality is a doomed enterprise as hopeless as folk physics.

# EXOSKELETONS

*(See also Prosthetics, Warbots)*

In biology, exoskeletons are the external skeletons that support and protect the bodies of all insects and certain animals—shellfish, for example. We humans have to design and build our own. For example, we've long designed and used armor to protect ourselves. Today, artificial exoskeletons promise to fulfill some of the fantasies of superhuman strength and invulnerability promised us in our comic books and heroes. People like this. Indeed, the single blog post that got the most page views for *h+* magazine was "I Am Iron Man," about "a cybernetic bodysuit that augments body movement and increases user strength by up to tenfold."

## THE HAL SUIT

The HAL Suit, which augments strength ten times, went on sale for $42,000 in 2009. It's manufactured by Cyberdyne—the company's name taken directly from the evil fictional corporation in *Terminator*. Real-life techies, particularly of the transhumanist sort, love to provoke by incorporating dystopian fantasies into their worlds. I've heard that DIY exoskeletons are a coming thing.

## More About HAL

**Tristan Gulliford:** The HAL (Hybrid Assistive Limb) suit works by detecting faint bioelectrical signals using pads placed on specific areas of the body. The pads move the HAL suit accordingly. The Cyberdyne website explains: "When a person attempts to move, nerve signals are sent from the brain to the muscles via motoneuron, moving the musculoskeletal system as a consequence. At this moment, very weak biosignals can be detected on the surface of the skin. HAL catches these signals through a sensor attached on the skin of the wearer. Based on the signals obtained, the power unit is controlled to wearer's daily activities."

Among the potential applications, Cyberdyne is emphasizing helping people with movement disabilities, augmenting strength for difficult industrial tasks, disaster rescue, and entertainment.

## Toward the Supersoldier

**Surfdaddy Orca:** Under contract from the Army, a team at Raytheon Sarcos, led by Stephen Jacobsen, built an exoskeleton called XOS. Looking something like Ripley wearing the industrial exo-suit power loader in the classic science fiction film *Aliens,* software engineer Rex Jameson used his XOS to run, jump, and even speedbox a punching bag. Jameson also was able to do a lengthy series of reps on a weight machine using two-hundred-pound weights. "He stopped because he got bored," Jacobsen says, "not because he was tired."

## DIY Exo

**Surfdaddy Orca:** Carlos Owens' eighteen-foot-tall, one-ton prototype flame-throwing "mecha" exoskeleton looks like something out of a *Transformers* movie.

Built over four years at a cost of $25,000, Carlos invented mechanical muscles to control the mecha's joints using twenty-seven hydraulic cylinders and a pump powered by an eighteen-horsepower engine. Like the character Ripley from *Aliens*, a driver climbs inside the exoskeleton to control the mecha's ability to walk, bend down, or open its hands. Steel cables transmit the driver's movements to the hydraulics. Capable of shooting flames from its hands, NewMech NMX01-1A would make an awesome foe at a demolition derby.

# EXTROPIANISM

*(See also Max More and Natasha Vita-More)*

The word "extropy" was conjured by some transhumanists in the 1980s to be the answer to—or opposite of—entropy. Entropy is the tendency of systems, both in nature and in culture, to run down, so extropy is the human will and ability to use intelligence and tools to improve and evolve better systems. Before the term "extropy," this was usually described as negentropy or negative entropy.

Formed by Max More and Natasha Vita-More at the end of the 1980s, the Extropy Institute was the first manifestation of organized transhumanism. Max More defined the principles of extropy as perpetual progress, self-transformation, practical optimism, intelligent technology, open society, information and democracy, and self-direction.

They immediately started publishing *Extropy: The Journal of Transhumanist Thought*. In 1991, they started the Extropians email list, a popular online watering hole for techno-dreamers. Extropians tended to be way far from mainstream reality visionary and libertarian geeks. On the Extropian List, they talked, argued, and fantasized about uploading their brains for immortality, using nanotechnology to make just about anything (or to smother the Earth in runaway self-replicating key lime pie), and, of course, overthrowing all governments everywhere.

## EXTROPIANS WERE THE EARLY ORGANIZED TRANSHUMANISTS

Throughout the early '90s, extropianism and transhumanism were virtually synonymous. To this day, some commentators, including the science fiction writer Charles Stross, satirize transhumanism and singularitarianism as though those movements were dominated by the same sorts that amused him on the Extropian List. In reality, those movements are now more diverse, both culturally and politically, but you can still find many examples of the extropian archetype.

> **Max More:** Organizational extropy is a crucial part of reaching a posthuman future. Human beings co-evolve with technology, and organizations are a social part of that technology. If we're to understand the past, present, and possible improved future of human beings, we must understand the range of potential organizational architectures and processes and their shaping factors.

# F.M. ESFANDIARY/
# F.M. 2030

F.M. Esfandiary—who changed his name to F.M. 2030—is generally recognized as the main philosophical progenitor of transhumanism. He first began using the terms "transhuman" and "posthuman" in 1966 in a course on the impact of technology on humans at The New School in New York City.

## MONDO BEYONDO

His 1973 book *Up-Wingers: A Futurist Manifesto* was built around the idea of humanity accelerating evolution via technology and advocated for most of the basic tropes of the movement: overcoming aging and death, living beyond Earth, limitless wealth, genetic enhancements, intelligence-increasing implants, and a species-wide, transparent, participatory human culture. Esfandiary also believed strongly in the detribalization of humans—that humans would move beyond their identifications with families, tribes, races, genders, and nation states.

Other important books by F.M. include *Telespheres* (1977) and *Are You a Transhuman?* (1989). Mr. 2030 died in 2000 and is now in cryonic suspension at Alcor.

# FRIENDLY AI

*(See also Artificial General Intelligence, The Singularity)*

Closely tied in to the notion of The Singularity, friendly AI presumes that we can create intelligences many times greater than our own and explores ways in which we can program them to be friendly to humans. Eliezer Yudkowsky developed the concept while at The Singularity Institute, where it became a central area for inquiry.

Science fiction writer Isaac Asimov imagined something like friendly AI with his Three Laws of Robotics, written in 1942. These are:

1. A robot may not injure a human being or, through inaction, allow a human being to come to harm.

2. A robot must obey the orders given to it by human beings, except where such orders would conflict with the First Law.

3. A robot must protect its own existence as long as such protection does not conflict with the First or Second Law.

Leading contemporary AI thinkers claim to have shown Asimov's laws to be flawed.

One problem not often mentioned by friendly AI thinkers is military research into intelligent machines. While a balance of terror may, arguably, be functional, it doesn't quite qualify as friendly.

## But Won't Somebody Somewhere Make Mean AIs?

**Hugo de Garis:** Eliezer Yudkowsky hopes that he and others will be able to mathematically prove that it is possible to design an intelligent machine that (of logical mathematical necessity, given its design) will be forced to remain human-friendly, even as it redesigns itself for increasing levels of intelligence . . . Even if friendly AI designs can be created, it does not automatically follow that they will be universally applied . . .

# FORESIGHT INSTITUTE

*(See also Nanotechnology)*

K. Eric Drexler introduced the concept of nanotechnology—working with matter at the scale of molecules—to the world at large with the release of his book *The Engines of Creation* (1986). He and Christine Peterson (then his wife) founded the Foresight Institute of Nanotechnology that same year. Although its charter was to guide technology and make it a positive force for humanity, the focus was nanotechnology.

Since Drexler and Peterson divorced, Christine has become the driving force behind the institute. Smart, dynamic, and well liked, she's one of the treasured sane people in the transhumanist movement.

**Surfdaddy Orca:** The Foresight Institute established the Feynman Grand Prize in 1996 to motivate scientists and engineers to design and construct a functioning nanoscale robotic arm with specific performance characteristics. This $250,000 incentive prize is somewhat analogous to the Loebner Prize for Artificial Intelligence (AI)—a Grail of sorts.

Either a human-level AI agent or a nanoscale robot arm will bring big changes.

The Foresight Institute is also helping to set the agenda for the beneficial applications of nanotechnology:

1. Providing Renewable Clean Energy

2. Supplying Clean Water Globally

3. Improving Health and Longevity

4. Healing and Preserving the Environment

5. Making Information Technology Available To All

6. Enabling Space Development.

# FUN THEORY

Leave it to Eliezer Yudkowsky to examine the scientific, mathematical, and programming challenges and potentials of fun. In a 2002 essay titled "Singularity Fun Theory," he asks such questions as: How much fun is there in the universe? What kind of emotional architecture is necessary to have fun? Does it require an exponentially greater amount of intelligence (computation) to create a linear increase in fun? Is self-awareness or self-modification incompatible with fun? Is (ahem) "the uncontrollability of emotions part of their essential charm?" It goes on from there to an honest (i.e. inconclusive) end.

Is the "Singularity Fun Theory" essay fun to read? Well, I think Sheldon Cooper would be beside himself (even without cloning). For this reader, well . . . yeah, it's mildly amusing.

## NEW FUN THEORY IS LESS FUN

In 2009, Yudkowsky wrote "The Fun Theory Sequence," which he says replaces his original essay. It's more complete and analytic, but less fun to read.

# GAMIFICATION

*(See also XPRIZE)*

In 1938, a Dutch cultural theorist, Johan Huizinga, invented the notion of the *Homo ludens* or "man at play." While Huizinga was just trying to prove that play was important and should be integrated into any culture, during the 1960s, the notion arose that life should be lived mostly at play, or even, if you were tripping, that everything should be experienced as play.

Gamification is sort of a compromised version of this idea for busy techno-times, one that is being integrated, at least a bit, into the corporate world. The idea is that problem solving and productive follow-through becomes more interesting, enjoyable, and creative if they're reimagined as a game.

Gamification can work on large scales. The XPRIZE has gamified the privatization of space and other big advances. As a gamification wiki puts it, "Gamificiation is the concept of applying game-design thinking to non-game applications to make them more fun and engaging." This can be enjoyable, or it can be absurd, as when companies try to force a happy face on bored workers who would rather be elsewhere.

Visionary architect Buckminster Fuller suggested using game playing to get people engaged in solving big social problems in 1961. He called the idea "World Game." Jane McGonigal has also explored it under the term "gamification" in her book *Reality Is Broken: Why Games Make Us Better and How They Can Change the World.*

**Surfdaddy Orca:** Dr. Byron Reeves—professor of communications at Stanford University and researcher into the psychology of media—has a prescription for boredom, repetition, and stress at work: turn work into play with online games.

With increased global competition, employee productivity and engagement having become more critical to businesses, Reeves argues that the user experience provided by game technology offers a "tantalizing solution" for business. This goes beyond online training tools. He advocates implementing elements of games such as World of Warcraft or virtual worlds such as Second Life to solve a host of business problems with "morale, communication, and alignment all while honing skills like data analysis . . ."

"Online games are very iterative," states a recent IBM report entitled "Virtual Worlds, Real Leaders." "Leadership happens quickly and easily in online games, often undertaken by otherwise reserved players, who surprise even themselves with their capabilities." Online games such as World of Warcraft can involve an overriding goal for a team of players—there are a series of raids or missions that make up the journey, each of which requires leadership of player groups of varying size. This gives many players the opportunity to "try on" leadership roles.

## Taxes Can Be Fun?

**David Helgason (interviewed by Ray Huling):** There's a tax-planning tool like this. They're competing with TurboTax and building game design into the product. It's funny, because it has to be the most boring field, but I mean, that's the point. You can make it slightly challenging and give people little reasons to sort of play these tax tools—beyond, you know, not going to prison!

# GENOMICS

*(See also Biohacking, George Church, Synthetic Biology, Telomeres)*

Genomics is the study of the complete set of DNA (the genome) within a cell. It's part of the larger field of genetics, which is the study of single genes. It is considered part of the "omics" revolution in biology, along with proteomics, metabolomics, and materiomics, which also study large sets of biological data, and epigenomics and metagenomics, which are omics inside of genomics.

## SEQUENCING, ASSEMBLY, AND ANNOTATION

Genomics has three parts. First, DNA of an organism is studied using one or more of the various forms of sequencing: shotgun, high-throughput, or Illumina dye. The order of the nucleotide bases is determined in small fragments of the DNA, because current methods cannot read entire genomes.

Then, the fragments are merged into a full DNA sequence, which can be either a reconstruction of the original DNA, or something new. Finally, the sequences are annotated with information about their structure and function.

In a noteworthy example of international collaboration, the Human Genome Project was declared a success in 2003, with 99 percent of the human genome mapped with 99.99 percent accuracy. (Because genomes are usually unique, multiple variations of each gene were sequenced.)

## MANY POSSIBILITIES

Virtually all human ailments involve our genes in one way or another, so the rapid growth in this knowledge is encouraging. As with many transhumanist technologies, costs are dropping rapidly. Medical researchers see hope for better diagnostic tools and treatments for major killers like cancer, diabetes, and cardiovascular disease, and for regenerative medicine. Biohackers and transhumanists see genomics as a tool for programming cells as if they were computers, for anti-aging interventions, creating bacteria for producing biofuels, and even creating entirely new living organisms and bioengineering humans for space travel.

## THE CRISPR BREAKTHROUGH

Genome editing took a huge leap forward in 2013 with the development of the Crispr technique. Previous methods of gene editing were clumsy and risky, inserting DNA into the genome at random. The Crispr technique is incredibly precise, able to edit individual nucleotides without messing up anything else. Plus, it's easy enough that novice lab assistants can do it. Suddenly, treating incurable viruses like HIV or fixing genetic defects in human embryos seems tantalizingly close, though laws against "designer babies" might stand in the way.

## GETTING PERSONAL

One of the most promising areas is personal genomics: using knowledge of your genes for personalized, predictive, and precision medicine. For example, pharmacogenomics tailors drug therapy to an individual's genetic make-up, because a drug or dosage that helps one person may do nothing for, or actually harm, someone else. However, there are some issues.

## TOO MUCH INFORMATION

With the cost of genome sequencing rapidly dropping, there's been exponential growth in databases of sequenced genomes. Researchers generated so much data that they sent discs by FedEx, though now file transfers are often done in the cloud. There are now many bioinformatics companies analyzing this deluge of data, but we are many years away from making sense of it all.

## GENETIC DETERMINISM, GENETIC RELATIVISM, OR GENETIC IRRELEVANCE?

Even if all that data could be perfectly understood and translated into clinical practices, it would be no panacea. Despite what you may have learned in biology class, the old dogma that genes control the processes of the cell turned out to be false. A cell can even live with its nucleus removed (though it will eventually die for lack of new proteins). Only about 5 percent of people have an abnormal gene that will create a major health issue. Diet, lifestyle, and environment are often far more important factors.

And then there's epigenetics. This rapidly growing area involves the ways that environmental factors such as diet can change gene expression, producing heritable changes that aren't explained by DNA alone. (Yes, Lamarck and Lysenko have been partially vindicated, sort of.)

## WE'RE FROM THE GOVERNMENT AND WE'RE HERE TO HELP YOU

The FDA regulates medicine and medical devices, and being risk-averse, has slowed the implementation of genomics-based medicine. Life-science companies are improving DNA sequencing faster than the FDA can handle it all. Even the software for interpreting the results of DNA tests for doctors must be approved. The FDA is also not thrilled with the idea of people taking medicine into their own hands, so personal genomics may face extra hurdles.

Still, genomics holds much promise, and is a crucial part of the transhumanist revolution.

## Optimistic About Genomics

**João Pedro (interviewed by Ben Goertzel):** I have a
of aging (or DevAge as I like to call it) . . . DevAge su
pects of aging follow pre-determined patterns encoded in
part of developmental processes. In other words, the genome does indu
contain instructions that drive aging. For anti-aging interventions this has
profound implications, because it means that if we can unravel the regulatory elements of these processes via genomics and bioinformatics, then
we may be able to manipulate them and therefore develop new anti-aging
interventions.

## No Longer Optimistic About Genomics

**Terry Grossman:** I used to be a big believer in the enormous potential of
genomics, and each of my two previous books with Ray Kurzweil, *Fantastic
Voyage* and *TRANSCEND: Nine Steps to Living Well Forever*, had chapters
devoted to this topic.

Ray Kurzweil is widely regarded as one of the world's foremost inventors
and futurists, and he has made predictions for what is likely to occur in the
future in the field of genomics. Yet, these days I find that I am feeling far less
confident about the near term prospects for this "promise."

Currently I have moved much closer to the idea of "genetic irrelevance,"
the idea that in the overwhelming majority of cases, our genes are of much
less importance in determining our fate and that the environment in which
we live and the lifestyle choices we make are of greater importance.

## Still Optimistic About Genomics: Ray Kurzweil Responds

**Ray Kurzweil:** There has been recent disappointment expressed in the progress in the field of genomics . . . The problem starts with the word "genomics." The word sounds like it refers to "all things having to do with genes." But as practiced, it deals almost exclusively with single genes and their ability to predict traits or conditions, which has always been a narrow concept. The idea of sequencing genes of an individual is even narrower and typically involves individual single-nucleotide polymorphisms (SNPs) which are variations in a single nucleotide (A, T, C, or G) within a gene, basically a two-bit alteration.

To put this narrow concept of genomics into perspective, think of genes as analogous to lines of code in a software program. If you examine a software program, you generally cannot assign each line of code to a property of the program. The lines of code work together in a complex way to produce a result. Now it is possible that in some circumstances you may be able to find one line of code that is faulty and improve the program's performance by fixing that one line or even by removing it. But such an approach would be incidental and accidental, it is not the way that one generally thinks of software. To understand the program you would need to understand the language it is written in and how the various lines interact with each other. In this analogy, a SNP would be comparable to a single letter within a single line (actually a quarter of one letter to be precise since a letter is usually represented by eight bits, and a nucleotide by two bits). You might be able to find a particularly critical letter in a software program, but again that is not a well-motivated approach.

The language of the genome is the three-dimensional properties and interaction of proteins. We started with individual genes as a reasonable place to start, but that was always going to be inherently limited if you consider my analogy above to the role of single lines in a software program . . .

To reverse-engineer biology we need to examine phenomena at different levels, especially looking at the role that proteins (which are coded for in the genome) play in biological processes.

# GEORGE CHURCH

*(See also Genomics, Synthetic Biology)*

George Church is Professor of Genetics at Harvard Medical School and a pioneer in genetic engineering, personal genomics, and synthetic biology. He developed the first methods for direct genome sequencing, codeveloped others, and his innovations have spurred the creation of a dozen companies in medical genomics and synthetic biology.

## AUTOMATION AND OPENNESS

Professor Church is perhaps best known for the Personal Genome Project (PGP), which aims to advance personal genomics by sequencing and publicizing the complete DNA sequences and medical records of 100,000 people. The genome sequencing at PGP is done by Polonator, an automated "next generation" sequencing machine he developed with his team. Polonator is less expensive to buy and operate than previous devices. Church is a big believer in open access, so all PGP data will be public domain, and the Polonator is open source hardware and software.

He and his team also developed DNA chip synthesizers and Multiplex Automated Genomic Engineering (MAGE), a tabletop machine for quickly and cheaply modifying genomes.

## NEW USES FOR DNA

Church has also co-invented DNA detectors for dark matter, and used DNA as a digital storage medium. His feat of cramming 700 terabytes into one gram of DNA means that, in theory, all human knowledge could be copied into a few tons of DNA that could be easily stored for thousands of years.

He's also part of the team behind the BRAIN Initiative (Brain Research through Advancing Innovative Neurotechnologies), which hopes to develop minimally invasive but high-resolution methods of detecting, recording, and manipulating neuronal activity.

## Speeding Up Biological Evolution

**Surfdaddy Orca:** Biological evolution as observed in the fossil record spans millions of years. Machine evolution, arguably the progeny of biological evolution in the form of ever more sophisticated human-generated computational devices and associated AI software, has spanned maybe 150 years since the landmark design of Charles Babbage's Difference Engine (the first mechanical "computer" that was never actually built but was designed during Darwin's lifetime). Now, with Church's genetic sequencing technology, biological evolution can occur in days.

This, in fact, is what Church and his team of researchers is now able to demonstrate. Called multiplex automated genomic engineering (MAGE), up to fifty changes to a bacterial genome can be made nearly simultaneously to accelerate development of bacterial cells.

# GRAPHENE

*(See also Nanotechnology)*

If you could unroll a carbon nanotube, you'd have a sheet of graphene: pure crystalline carbon, with the atoms arranged in a hexagonal (honeycomb) lattice just one atom thick. It was a hypothetical material until Andre Geim and Konstantin Novoselov at the University of Manchester in the UK produced some in 2004, which led to their 2010 Nobel Prize in physics.

## EXCEPTIONAL PHYSICAL, CHEMICAL, AND MECHANICAL PROPERTIES

Graphene is one of the most exciting recent developments in materials science and nanotechnology. It's the strongest material known, conducts heat and electricity better than copper, is extremely light (0.77 grams per square meter) and nearly transparent, and yet so dense that even helium cannot pass through it.

## TOO MANY USES TO LIST

It can make metal composites hundreds of times stronger. Mixed into plastics, it can make them stronger, electrically conductive, and heat resistant. Graphene oxide is excellent at absorbing radioactive waste. It shows promise in rustproofing coatings, detecting explosives, and for creating artificial muscles, better bulletproof vests, and improved night-vision goggles.

Graphene is used in disease diagnostics, drug and gene delivery, and photothermal therapy for Alzheimer's and cancer. It can sequence DNA more cheaply by filtering molecules by size.

Lockheed Martin has produced a perforated form they call Perforene. Perforene promises to make desalination much cheaper because it requires much less pressure to operate than standard reverse osmosis filters. Fresh water shortages may become a thing of the past.

Perhaps its greatest promise lies in electronics: improved (even flexible) flatscreen displays, superconducting cables, optical data storage, and thin speakers. Graphene photodetectors integrated into silicon chips are smaller and faster, and could improve fiber optic cables and revolutionize optoelectronics. A 3D version of graphene may replace platinum in dye-sensitized solar cells, making them much cheaper. Graphene supercapacitors could replace batteries, survive more charging cycles, and power laptops for days.

## SAVING MOORE'S LAW

Silicon transistor densities (and thus processing speed and memory capacity) have been doubling about every two years for over forty years, which is why it's possible to have a phone in your pocket with more computing power than it took to land on the Moon. But silicon transistors can only get so small before they hit the wall of quantum physics, and that now looks to happen within a decade.

The electronics industry has wet dreams of graphene CPUs hundreds of times faster than today's silicon, but its excellent conductivity gets in the way. Silicon is a semiconductor, with a small band gap, so you can turn it on and off, which means you can make transistors out of it. But graphene has no band gap, so what to do?

So far no one has been able to produce a good artificial band gap in graphene, but Guanxiong Liu and others at the University of California (Riverside) have designed a graphene transistor that replaces the band gap with negative differential resistance. There's also a related material called carbyne, a string of single carbon atoms with twice the tensile strength of graphene. Carbyne has a band gap that can be altered by stretching or twisting. If one of those (or something similar) works out, Moore's Law can just shift gears and keep going.

If graphene were a stock, it would be a long-term buy.

# GRINDERS

*(See also Biohacking, Citizen Scientists, Cyborgs, Quantified Self, Synthetic Biology)*

## SELF-CYBORGIZATION WITH KITCHEN KNIVES

In computer gaming, grinding refers to repetitive actions that improve your character or unlock features, but the term "grinders" may have been first linked to extreme self-biohackers in *Doktor Sleepless*, a transhumanist science fiction comic book series written by Warren Ellis.

Grinders are self-identified biopunks noted for their hardcore, underground, ready-for-anything attitudes. They practice functional and extreme body modification that frequently involves DIY surgery. A popular modification is to implant magnets in fingertips, allowing them to feel electromagnetic fields. One tattoo artist implanted magnets to hold his iPod Nano. Other grinders have experimented with implanting RFID tags. Another project discussed on the grinder discussion board Biohack.me is to try replacing dietary vitamin A with porphyropsin (vitamin A2), in hopes of expanding vision into the near-infrared.

## WHAT WOULD YOU LIKE TO BE TODAY?

Grindhouse Wetware is a leading group of grinders, operating under a mutated version of the old Microsoft slogan. One member, Tim Cannon, recently had the Circadia device (née HELEDD) implanted in his arm. No surgeon would do it, so they found a piercing artist, who could only use ice as an anesthetic. The first version of Circadia is relatively primitive and awkwardly large (slightly smaller than a deck of cards). It records Cannon's body temperature and uses Bluetooth to send the data to his cell phone. Future versions will read and save a wider range of biomedical data. Grindhouse Wetware is also working on Bottlenose (which allows you to feel sonar, UV, WiFi, and thermal data) and Thinking Cap (for brain stimulation).

A note to the FDA: these are all *art projects*, really, so please don't consider them medical devices and try to regulate them. Grinders can't afford that. Heck, they often can't afford real scalpels, and just sharpen and sterilize kitchen knives.

## EVERLASTING GRINDER LOVE

Nick Bostrom has talked about biohacking not just the physical, but the emotional. Genetic modification of prairie voles has transformed them from polygamous to monogamous, so perhaps a biohacked hormone could make your love truly everlasting. If that happens, grinders will have made new enemies among divorce lawyers.

**Lepht Anonym:** I'm sort of inured to pain by this point. Anesthetic is illegal for people like me, so we learn to live without it; I've made scalpel incisions in my hands, pushed five-millimeter diameter needles through my skin, and once used a vegetable knife to carve a cavity into the tip of my index finger. I'm an idiot, but I'm an idiot working in the name of progress: I'm Lepht Anonym, scrapheap transhumanist. I work with what I can get.

Sadly, they don't do it like that on TV. The art of improving the human is shiny and bright in the media. You see million-euro cryogenics policies and hormonal life-extension regimes that only the elite can afford. You see the hypothesis of an immortal silicon body to house your artificially enhanced mind. You could buy that too, maybe, if you sold most of your organic body and the home it lives in. But you can do something to bring it down a notch: homebrewing.

My first foray was into RFID (radio frequency identification). After one visit to an outraged state GP here in Scotland ("I wouldn't do it even if I could, and I have no idea why you want to do it!"), I was fairly certain I'd been born in the wrong country for that—here, doctors would be struck off the records for helping me. I was on my own.

Luckily, I'm far too stupid to be stopped by bureaucracy. I bought my first Swann-Morton scalpel online, scrubbed the cleanest bathroom we could get with household bleach, settled myself crosslegged over the bathtub with my spotter, and poised the blade over the Biro-ink line I'd drawn for guidance. For a few minutes, I doubted whether I'd even be able to do it—cutting yourself open is not something we're adapted to be good at. Contemplating St. Gibson, I took the plunge.

It took a few weeks to heal, and when it did, with some help from my local gurus I was able to program a cheap open-source Phidgets RFID reader to recognize the chip's hexadecimal ID. The piece of C code that did it resided on a Linux machine and ran in the background while the reader was connected, waiting for my chip to show up. In short, it could see me and print a little "hi" when it did.

That's just garbage programming, too—you can see the potential if it were given to a real coder. The chip works with any homebrew RFID project. You want a laptop tracking system? A door that only lets you in? A safe that won't allow keypad input if you're not next to it? All you need is an ampoule (you get five for a euro, the last time I checked), from any RFID hobby place, a cheap reader, and a touch of disregard for risks. Salvage a keyboard from your local dump and you've got a simple system for bioidentification.

The implants sit in various places under my skin: middle fingertips of my left hand, back of the right hand, right forearm—tiny magnets, five or six millimeters across, coated in gold and then in silicon to isolate the delicate metal from the destructive environment of your body. They're something of an investment at about thirty euros apiece, and hard to get hold of, but worth pursuing. When implanted, they become technological sensory organs.

There's an entire world of electromagnetic radiation out there, invisible to most. Our cities are saturated with it. A radio, for instance, gives off a field that's bigger than the device itself. So do power supplies and wires in the walls. The implants pick up on the fields, and because they're magnets, they fizz with gentle electricity, telling you this hard drive is currently active, that one is turned off, there's the main line in the wall . . .

Watching commercials for vitamin pills on TV and thinking you need a mad scientist's lab to be a transhumanist? You don't. I've got no money, talent, or backing. You just need curiosity and the willingness to withstand some pain. Risk, not money, is our obstacle.

Turn off the TV. Pick up that needle. Come to the junkyard.

# HUMANITY+

Originally called the World Transhumanist Association, the organization Humanity+ started off as an attempt to ditch the transhumanism label, mainly because some members of the WTA believed that it was associated with extreme libertarianism, excessive utopianism, and an underlying nihilism towards ordinary humanity. They hoped for a somewhat more sedate and considered discourse that emphasized the humanistic aspect of the transhumanism idea. One way they distinguished themselves for a time from these earlier transhumanist traits was to elect a socialist, James Hughes, to be the Executive Director.

Over the years, Humanity+ has changed leadership many times, and the political leanings of those elected have been varied, but it has continued to be characterized by a wide variety of views and concerns.

## THE EVER RESPECTABLE R.U. SIRIUS PLAN

In 2008, I was asked by then-director James Clement (an economic centrist and cultural libertarian, incidentally) to edit a new electronic magazine for the organization titled *h+*. While the irony of hiring someone named R.U. Sirius to bring a more respectable image to transhumanism may have only been noticed by one person (me), I did my best to organize an interesting, smart, exciting PDF and online flipbook magazine for them, and I believe that I succeeded. However, the sensibility didn't really exist in this fast-forward milieu to appreciate the magazine as a magazine, and most people just read the articles when we ran them as blog posts. The blog became quite popular and so the magazine concept was dropped.

## BROKE-ASS MEMBERS OF THE RULING ELITE

Occasionally, one will read a conspiracy theory that imagines that the Humanity+ organization is a well-funded tool of whatever version of the New World Order the author has conjured up. In fact, the organization could never really afford a single full-time employee, and after briefly turning the magazine over to a company called Better Humans— also involving James Clement—they regained control of the website. Currently, Peter Rothman does a fine job of editing the blog.

Despite its low funding, Humanity+ still manages to put on excellent and well-attended conferences. Their current chairperson is Natasha Vita-More, someone closely associated with the old transhumanist tendencies that they were supposedly trying to escape.

According to the Humanity+ website, "Approximately 6,000 people belong to Humanity+ from more than 100 countries, from Afghanistan to Brazil to Egypt to the Philippines." However, the idea that Humanity+ could replace transhumanism as a brand identity has long since been abandoned.

**R.U. Sirius:** I read an article, a rant really, in *Counterpunch* titled "If Only Glenn Beck Were a Cyborg," in which author David Correia goes to the recent Singularity Summit and finds that the "Singularity Movement" is "made up of university scientists, technocapitalists, and military funders" and is a tool of "the bourgeois dream of class domination and faith in technoscience," and that it hides "its corporate face with scientific fact, military control with techno-enthusiasm, and ruling class ideology for general human benefit."

So there I am curiously reading along, when I come to this sentence in the third paragraph: "The journal, H+ (Humanity Plus), along with a host of institutions, centers and even universities advance a research agenda many refer to as 'transhumanism' in which the problems of human intelligence and immortality are primary subjects." And for a moment, I wondered, *What is this H+ Journal?* And then it hit me. He's talking about us! Us? A journal?! A tool of the ruling class yet. Does that mean my all-volunteer staff can finally get paid? I'm chuffed! Jeeves, fetch my saddle and bring Mr. Ed out to the courtyard. It's time to celebrate with a round of polo!

Well okay, getting more serious . . . arguably something like The Singularity could permanently stamp its corporate sponsors' logos on the exo-neocortex copies of us all as we upload our psyches onto slick little Apple devices where we will only get to access realities that have signed an exclusive contract with Steve Jobs' preserved data. And arguably, aspects of the Singularity movement may play into the fantasies of a few billionaires who have experienced a revolution in rising expectations over the last three decades . . . "Let's see, I've got the fourteen houses, the eight boats, the ten private jets, the private island in the Pacific, the surgically sculpted trophy wife . . . what else is there? I know. I'll live forever in a superbody on the Moon with Richard Branson!" The economics of all this are a lot more complex than billionaires plotting to live forever and screw the rest of ya, but I'll leave that for another time.

Correia, of course, paints a diverse culture with a broad brush—failing to distinguish between singularitarianism and transhumanism, for one, and amongst singularitarians for another. And what about this idea that The Singularity Movement is "made up of university scientists, technocapitalists, and military funders." Absolutely true. And hackers, open source fanatics, citizen scientists, ordinary tech enthusiasts, and dreamers of various political bents (including leftists). In other words, it has basically the same social and cultural dynamic that has been in play in technoculture since the mid-1980s.

# IMPLANTS

*(See also Artificial Hippocampus, Biohacking, Citizen Scientist, Cognitive Enhancement, Cyborgs, Grinders, Neurotechnology, Optogenetics, Prosthetics)*

Implants have been around for a while. The first heart pacemaker was implanted in 1958, and medical implants are now commonplace for repairing or replacing teeth, heart valves, or joints. Many pets now get implanted with RFID tags, and some body modifiers have gone beyond tattoos and piercings to implant horns and other shapes of silicone, Teflon, stainless steel, or titanium.

## GETTING CHIPPED

Biohackers have implanted chips to open doors and switch on lights. Attempts have been made to commercialize similar chips for identification, medical records, security, or tracking purposes, but none have proven successful, due to medical, security, and legal issues. Grinders are experimenting with implanting devices that monitor biomedical data.

## FUTURE POSSIBILITIES

Advances in medical research, nanotechnology, and digital technologies all indicate that implants will be an important transhumanist tech. Implants that dispense drugs are already undergoing FDA trials. So-called brain-machine-brain interfaces are being tested for treatment of strokes and other traumatic brain injuries. They work by taking signals from one part of the brain, processing them, and then stimulating another part of the brain that has lost its connectivity, bypassing the damaged area. Implanted brain-computer interfaces that allow full net access, as imagined in the *Intelligence* TV show, seem inevitable.

# IN VITRO MEAT

*(See also Cloning)*

The notion that tissue cultures could be developed into veritable animal flesh without the necessity of raising and slaughtering living creatures has been in circulation among tech enthusiasts for several years. Fans of classic science fiction will think of the giant mass of meat called Chicken Little in Pohl and Kornbluth's 1952 novel *The Space Merchants*. With current off-the-shelf biotechnology, it's become possible to grow edible meat in laboratory vats, starting from a single cell.

## BUT . . . I MEAN, *YUCK!*

In 2008, this idea got a boost from PETA (People for the Ethical Treatment of Animals). The animal rights group offered a $1 million prize for "the first person to come up with a method to produce commercially viable quantities of in vitro meat at competitive prices by 2012." The challenge was controversial among PETA supporters because . . . well . . . like, I mean . . . *yuck!* No one picked up the check. I'm pretty sure most people would find guilt-free, tasty meat eating is a yummy idea.

## THE WORLD'S MOST EXPENSIVE BURGER

The first in vitro produced meat, a $325,000 hamburger, was reported in *The New York Times* on May 12, 2013. According to the *Times* article, it was "assembled from tiny bits of beef muscle tissue grown in a laboratory and to be cooked and eaten at an event in London." At the taste test in London, the burger got mediocre reviews, with one of the tasters rationalizing, "It's close to meat." Surely, the taste can be improved . . . but can it be done without fat?

**Hank Pellissier:** In vitro meat—a.k.a. tank steak, sci-fi sausage, petri pork, beaker bacon, Frankenburger, vat-grown veal, laboratory lamb, synthetic shmeat, trans-ham, factory filet, test tube tuna, cultured chicken, or any other moniker that can seduce the shopper's stomach—will appear in a few years as a cheaper, healthier, "greener" protein that's easily manufactured in a metropolis. Its entry will be enormous; not just food-huge like curry rippling through London in the 1970s or colonized tomatoes teaming up with pasta in early 1800s Italy. No. Bigger. In vitro meat will be socially transformative, like automobiles, cinema, vaccines.

## Bye-Bye Ranches

**Hank Pellissier:** When in vitro meat (IVM) is cheaper than meat on the hoof or claw, no one will buy the undercut opponent. Slow-grown red meat and poultry will vanish from the marketplace, similar to whale oil's flame out when kerosene outshone it in the 1870s. Predictors believe that IVM will sell for half the cost of its murdered rivals. This will grind the $2 trillion global live-meat industry to a halt (500 billion pounds of meat are gobbled annually; this is expected to double by 2050). Bloody sentimentality will keep the slaughterhouses briefly busy as ranchers quick-kill their inventory before it becomes worthless, but soon Wall Street will be awash in unwanted pork bellies.

Special note: IVM sales will be aided by continued outbreaks from filthy, overcrowded farms—diseases like swine flu, mad cow disease, avian flu, tuberculosis, brucellosis, and other animal-to-human plagues. Public hysteria will demand preemptive annihilation of the enormous herds and flocks where deadly pathogens form, after safe IVM protein is available.

## Healthier Planet

**Hank Pellissier:** Today's meat industry is a brutal fart in the face of Gaia. A recent Worldwatch Institute report ("Livestock and Climate Change") accuses the world's 1.5 billion livestock of responsibility for 51 percent of all human (or cow) caused greenhouse gas emissions. Statistics are truly shitty: cattle crap 130 times more volume than a human, creating 64 million tons of

n the United States that's often flushed down the Mississippi River
h and coral in the Gulf of Mexico. Pigs are equally putrid. There's
arm in Utah that oozes a bigger turd total than the entire city of Los
s. Livestock burps and farts are equally odious and ozone-destroying.
Sixty-eight percent of the ammonia in the world is caused by livestock (creating acid rain), 65 percent of the nitrous oxide, 37 percent of the methane,
9 percent of the $CO_2$, plus 100 other polluting gases. Big meat animals
waste valuable land—80 percent of Amazon deforestation is for beef ranch-
ing, clear-cutting a Belgium-sized patch every year. Water is prodigiously
gulped—15,000 liters of $H_2O$ produces just one kilogram of beef. Forty percent of the world's cereals are devoured by livestock. This scenario is clearly
unsustainable, and in vitro meat is the sensible alternative. Once we get over
the fact that IVM is oddly disembodied, we'll be thankful that it doesn't shit,
burp, fart, eat, overgraze, drink, bleed, or scream in pain.

## Exotic and Kinky Cuisine

In vitro meat will be fashioned from any creature, not just domestics that
were affordable to farm. Yes, *any animal*, even rare beasts like snow leopards
or Komodo dragons. We will want to taste them all. Some researchers believe
we will also be able to create IVM using the DNA of extinct beasts—obviously,
"DinoBurgers" will be served at every six-year-old boy's birthday party.

Humans are animals, so every hipster will try cannibalism. Perhaps we'll
just eat people we don't like, as author Iain M. Banks predicted in his short
story, "The State of the Art" with diners feasting on "stewed Idi Amin." But I
imagine passionate lovers literally eating each other, growing sausages
from their comingled tissues overnight in tabletop appliances similar to
bread-making machines. And, of course, masturbatory gourmands will simply
gobble their own meat.

# INSTITUTE FOR ETHICS
# AND EMERGING TECHNOLOGY
# (IEET)

The IEET is the even more sober and socially responsible version of transhumanism to emerge since the wild and woolly days of the extropians. Transhumanist socialist James Hughes and British transhumanist philosopher Nick Bostrom started the nonprofit. According to their website, the Institute is "a center for voices arguing for a responsible, constructive, ethical approach to the most powerful emerging technologies." In the transhumanist tradition, their focus remains on the impact of human enhancement technology. Politically, they endorse the United Nations' Universal Declaration of Human Rights, generally a left-liberal interpretation of the human rights concept, making them diametrically opposite to the original libertarian trend in transhumanism. They have been accused of being a wee bit boring.

**George Dvorsky:** Over the past several years a good number of "futurists" and all-out naysayers have systematically worked to undermine and dismiss the potential for radical change to occur in the not-too-distant future. A number of commentators—including some of my colleagues at the Institute for Ethics and Emerging Technologies—have openly rejected the potential for paradigmatic changes to occur.

It's suddenly become very fashionable to pooh-pooh or sweep aside the pending impacts of such things as the looming robotics and manufacturing revolutions, the rise of super AI, radical life extension, or the migration of humans to postbiological form.

# LAW OF ACCELERATING RETURNS

*(See also Moore's Law)*

## EXPONENTIAL PROGRESS

This term was coined by Ray Kurzweil in his book *The Age of Spiritual Machines* (1999). He proposes that the rate of progress in technological and many other evolutionary systems tends to increase exponentially: a Moore's Law for everything.

## MORE KNOWLEDGE MEANS MORE INTERCONNECTIONS

As mentioned earlier, the idea of progress is a relatively recent historical development, and despite some pushback from such thinkers as Malthus, Nietzsche, Spengler, and Kirkpatrick Sale, the concept remains embedded in our culture. In the 20th century, the idea expanded. Technological progress came to be seen as not merely linear, but ever-accelerating. People like Robert Metcalfe saw it in the specific case of the value of telecommunication networks (Metcalfe's Law: the more people there are who have a device like a telephone, the more valuable each one becomes). R. Buckminster Fuller, Stanislaw Ulam, John von Neumann, Vernor Vinge, and Ray Kurzweil saw it as a more general trend.

It is perhaps most visible in the realm of pure information: scientific papers, patent applications, and web pages are all increasing at an accelerating rate. It's not simply that new information is being added to the total, but that each bit creates new connections with what is already there, both inside a given field and in others. Even if the number of information units in a given area isn't growing exponentially, their interconnections are.

## MORE INTERCONNECTIONS MEANS FASTER PROGRESS

Acceleration in knowledge leads to accelerated progress in the physical and social worlds. Louis Pasteur's discoveries led to healthier children, safer food, and richer societies. Physics papers lead to better computer chips, which lead to better phones and telecommunications, which then lead to MOOCs (massive open online courses), smartphone apps for diagnosing disease, and Facebook for sharing cute pictures of puppies and kittens.

It took generations for inventions like the telephone and the automobile to become widespread, but in a few decades mobile phones went from being expensive status symbols for the rich to something owned by every Third World goatherd.

Call it positive feedback (here, the good kind of feedback), a network effect, or even a bandwagon effect—the more you have, the more valuable each piece, and everything else, becomes. When the lines on your graph continually curve upward, they eventually become vertical: The Singularity.


## CRITICISMS

As they say, there's nothing riskier to predict than the future, and past performance is not a guarantee of future results. Individual technologies often follow an S-curve, like the left half of a bell curve. In the middle they look like they might go exponential, but then they flatten out (though Kurzweil argues that exponential progress is often in the sum of individual S-curves).

Look at aviation: the fifty-five years between the 1903 Wright Flyer and the 1958 Boeing 707 made aviation progress seem pretty exponential for a while, but the next fifty-five years blew that idea out of the sky. Today's big airliners are larger and more fuel-efficient, but air travel isn't any faster or more comfortable. Plus, today you have to pay extra for your airline meal and your checked bags, seats are smaller, and the threat of terrorists makes airports a huge, Orwellian hassle. Those atomic-powered, computer-piloted personal helicopters didn't show up on schedule. From a passenger's point of view, progress in air travel peaked decades ago.

Transhumanists believe in (or at least, are hoping for) a future that's more like smartphones and less like the TSA.

**Vernor Vinge:** When I began writing science fiction in the middle '60s, it seemed very easy to find ideas that took decades to percolate into the cultural consciousness; now the lead time seems more like eighteen months.

**Ray Kurzweil:** We won't experience one hundred years of progress in the 21st century—it will be more like 20,000 years of progress (at today's rate) because the pace of technological change is exponential.

# LIBERTARIAN
# TRANSHUMANI

*(See also Extropianism, Peter Thi*

There are many different political views in the world of transhumanism. If you look at a 2012 IEET poll, which cast a very broad net by inviting participants from the virtual places where transhumanists gather, you will find that when asked "what is your political position?" the majority of transhumanists fall into categories that are usually defined as "progressive" or "left." If you total up liberal, socialist, and Marxist selections, you wind up at roughly 54 percent. Libertarians come in at about 27 percent, a 2:1 ratio. (The remainder describe themselves as moderates.) Still, libertarianism and transhumanism are frequently paired in the minds of many critics and adherents.

## A LIBERTARIANISM OF THE BODY

As advocates of the individual's right to enhance and alter themselves, transhumanism is inherently libertarian regarding the body. Differences amongst transhumanists arise around what sort of social and economic relations should govern. Or whether, in fact, there should be any governing whatsoever.

## SOME LIBERTARIANS ARE LOUDER OR HAVE MORE MONEY

Despite the fact that libertarians no longer dominate transhumanism numerically, a number of prominent libertarians are vocal advocates and deep-pocket funders of transhumanist efforts. Most famous (or infamous) among them is Peter Thiel, who made his billion or so primarily as co-founder and former CEO of PayPal. The popular libertarian blogger Glenn Reynolds is a noted advocate. The libertarian magazine

is also notably transhumanism-friendly, particularly their science
or, Ronald Bailey. Recently, Zoltan Istvan's novel *The Transhumanist
Wager* presented a radically libertarian vision that has been compared to
that of Ayn Rand (Randians insist she wasn't a libertarian).

A number of transhumanists call themselves anarcho-capitalists, a
close and more extreme cousin of the libertarians.

**Malcolm McLuhan:** . . . A very European take on US technoculture is "I see
libertarian people!"

# LONGEVITY/IMMORTALITY

(See also Aubrey de Grey, Caloric Restriction, Cryonics,
The Methuselarity, Mind Uploading, Nutraceuticals, Telomeres)

Living beyond the perceived limits of an individual human life seems to
be the central obsession of transhumanist culture. Transhumanists are
against death. Many of the projects and developments discussed in this
book are explicitly aimed at the defeat of the Grim Reaper. If you think
death is okay, a transhumanist might call you a deathist. A deathist is an
enemy of transhumanism, just as a capitalist is an enemy of communism,
or a Marxist is an enemy of capitalism.

## BEGINNINGS

The quest for extreme longevity through contemporary science method
began to gather steam in the latter half of the 20th century. In 1962,
American physics professor Robert Ettinger proposed that the biological sciences would someday soon find a key to immortality, and that
cryonic preservation was a way that a person living in the 20th century
could keep himself intact until that great day. In 1964 he published *The
Prospects for Immortality*, the first popular call to arms for using technology against death.

The book was a hit, particularly with the major newspapers and
magazines of the time. Ettinger also appeared on various radio and TV
talk shows, including Johnny Carson's *The Tonight Show*. This was the
beginning of the movement for scientific immortality. By the late 1970s,
a number of cryonics companies had sprung up and a number of public
figures were talking about the prospects for immortality, among them
Durk Pearson and Sandy Shaw, Timothy Leary, F.M. Esfandiary, and
any number of science fiction writers including Arthur C. Clarke. The
pop science magazine, *OMNI*, first published in 1978, featured frequent
articles about the pursuit of extreme longevity.

## HEALTH EXTENSION

Many people object that they wouldn't want to live longer like an old person, but the notion of longevity or life extension is best understood as *health* extension. Many physical ailments arise out of aging, and interventions to slow, stop, or even reverse aging will be aimed at maintaining health, perhaps even before they are aimed at aging itself. The transhumanist movement is particularly dedicated not only to maximizing basic health, but in stretching our notion of basic health beyond its ordinary limits. Instead of just not being injured or diseased, transhumanists would have us feel vigorous, physically and mentally hyper-enabled, and very, very good.

## PROGRESS?

For many years, longevity fanatics searched for a particular mechanism—a "biological clock"—that could be monkey-wrenched. Many optimists predicted a longevity pill before the 21st century, based on the notion that there would be a relatively simple mechanism at work.

These days, many longevists now see it as a more complex process of fixing specific problems associated with aging. On the other hand, the experiments conducted by Cynthia Kenyon showed that radically extending the lifespan of at least some biological creatures is possible with a single gene mutation. She got some worms to live to twice their normal apparent biological limit, and now has a company (Elixir Pharmaceuticals) that is attempting to make that long-hoped-for longevity pill.

## ENTER GOOGLE

In 2013, radical life extension went *way* mainstream when Google became the major investor behind Calico, a biotech company with the specific agenda to end or slow aging and diseases associated with it. Art Levinson, who is also Chairman of Apple and Genentech, is the major mover and shaker behind the company. Cynthia Kenyon, who you just learned about, will be part of the team. (Is Google starting to get a little scary and exciting or what?)

## POLITICS

The pursuit of extreme longevity or even immortality is not exactly a populist cause. Most humans are just trying to survive and to have ordinary lifespans that aren't limited by scarcity and blighted by physical and emotional pain. And, as we've already noted, there is a popular suspicion that extreme longevity is—and will be—mostly for the wealthy. Nevertheless, some transhumanists—who either have great confidence in their persuasiveness or are particularly tone deaf— formed Longevity Parties. These political parties would have their societies put more money and effort toward the pursuit of longer lives. The Longevity Party is active (as far as we know) in the US, Canada, UK, Ireland, the Philippines, and the Netherlands.

A spinoff from the Longevity Party, The Longevity Alliance is engaged in petitioning organizations, including the United Nations, for changes in regulatory schemes for biomedical interventions on aging as well as the increase in funding for life-extension science. They have chapters in at least fifty countries.

### How Can We Get Rid of Death and Disease?

**Ben Goertzel:** The big question is when.

One of the main lessons we've learned in the last few decades, as our exponentially improving instrumentation has enabled us to gather more and more data about humans and other organisms, is that human bodies are damn complex. This is why, over a decade past the completion of the Human Genome Project, our knowledge of the human genome hasn't yet led to a parade of new blockbuster drugs. The complex self-organizing networks via which genes help produce organisms are something we're just now beginning to understand. And the human mind, powerful as it is, has a limited capability to comprehend complexity.

This gives rise to the question: Can the human mind, aided only by tools that have dramatically less general intelligence than itself, comprehend the human body? Or will it need help—perhaps from advanced AI programs, specialized for the task?

In 2005, I was invited to Australia to address a conference of geneticists, and over dinner I asked a group of them the following question: When do you think human biology will be solved?

Few of them had thought about the question before, but nearly all agreed it was a sensible one. If our current scientific paradigms hold reasonably true, then it seems likely that, in the same sense that we now understand basically everything we need to know about the physics of an automobile engine or a dishwasher, one day we will have pragmatically near-complete knowledge about human bodies.

Once the question had sunk in, answers I received ranged from "fifty years" to "thousands of years." I'm also well aware of the difficulties involved in cracking the life code. Still, I think thirty to fifty years is probably the right answer because I believe we're likely to create AI biologists that will romp and play where the human mind struggles.

## Living a Million Years . . . Give or Take a Few

**Reason:** The million year lifespan is not an unsupported pipe dream. Living for a million years is a goal that can be envisaged in some detail today: the steps from here to there laid out, the necessary research and development plans outlined, and the whole considered within the framework of what is permissible under the laws of physics, and what the research community believes can be achieved within the next twenty, fifty, or one hundred years.

Biotechnology is the first necessary step on this road of a million years. The biotechnology revolution, still in its early years, is a gateway to the future insofar as it will enable us to extend our healthy lifespans by repairing the evolved world of nanoscale machinery within our cells and other vital biological systems.

The future is only golden for you and I personally if we live to see it. This golden future is one in which our biochemistry does our bidding, aging can be repaired, and molecular manufacturing is in full swing. It will be an age of bioartificial bodies, minds transferred to new and more robust mechanisms, strong artificial intelligences, an end to most scarcities, and, indeed, anything you might imagine that the laws of physics permit and enough time has passed to develop.

If you project out based on accident rates today, you'll find that an age-less human sustained by biotechnologies of cellular and biochemical repair has a life expectancy in the range of one to five thousand years. Sooner or later that piano is going to fall upon your head hard enough that even advanced medical technology cannot fix your injuries in time. So the million year lifespan—how could that be achieved? The short and not terribly infor-mative answer is that it will be accomplished by using advancing technology to dramatically reduce your vulnerability to fatal accidents, murder, and other unfortunate events that produce the same outcome. Once you start looking at living for even 100,000 years in much the same shape as you are today, it becomes apparent that almost any activity bears an unavoidable minimum level of risk that will jump up and kill you. Eating, swimming, walking, breath-ing. Stretch out the time frame far enough and the improbable and fatal will eventually happen.

The way past these risks is to change your form: your risk of fatality for any given activity is a function of your human physiology. Once the research and development community has achieved the goal of practical biotechnol-ogies for the repair and reversal of aging, that will give us all a few hundred years of life in comparative statistical safety. Technological progress will continue across that long period of time, and I can't imagine that much of the toolkit needed for the next step in long-term risk reduction will remain beyond the human civilizations of the 2200s. I would get my neurons re-placed—slowly, one at a time over time, to ensure continuity of the self—with some form of much more robust, easily maintained nanoscale machinery. That allows these engineering possibilities:

- Swapping out the body for whatever machinery of transport and sup-port best minimizes risk
- Moving most of the business of life into a virtual world
- Physically separating my neurons while still remaining alive, con-scious, and active.

Physical distribution of the self across many disparate locations is the key point when it comes to considering risk over the long term. Locations have much the same issues with time, probability, and bad events as people do. Meteorites happen, as do landslides, earthquakes, war, and volcanoes. The way to reduce your location-based risk dramatically is to spread out.

What to do about all of the astrophysical and grand geological risks? Spreading out is an option: boost up the size of your vehicles and neuron machines to shrug off the worst case radiation projections for a nearby supernova. Provide them with the means to move about the solar system, and become a spacefaring entity, spread out over a sizable selection of orbits. By that point in time, your physical presence resembles a small country of machinery, automation, and layers of delegation: perhaps you are a million heavily shielded self-powered containers and transmission systems distributed beyond Pluto's orbit.

## For Rich People Only?

**R.U. Sirius:** As serious life extension appears on an ever nearer horizon, this may be a good time to entertain current anxieties about it while thinking beyond the two extant competing simplistic arguments.

Among the current conflicting views we find these:

**A:** Hyperlongevity will be for rich people only.

**B:** Technologies get distributed to more and more people at an increasing rate of speed through the auspices of the free market. Demand increases. Production increases. The price gets lower. Demand increases. Production increases. The price gets lower . . . ad infinitum. In fact, the wealthy who are the early adopters of a new technology get to spend a lot of money on crappy versions of new technologies that are not ready for prime time.

At the risk of being obvious, it seems like there's a lot of room in the middle for more nuanced, less certain views.

Let's start with the notion that hyperlongevity is only for the super rich. While I largely buy the argument that important technologies get distributed to huge numbers of people via the market, it's never a good idea to indulge in technological determinism (unless your livelihood depends on making sanguine pronouncements about the future). For one instance, if hyperlongevity depends on surgical intervention, say, for organ replacement, and if it remains dependent in that way for a long time, hyperlongevity could, in fact, turn out to be only available to the wealthy. Indeed, organ replacement already stands as a longevity enhancement that is not very well distributed. Additionally, anyone who has had AIDS for a few decades can tell you that the drug combinations needed to maintain a semblance of health remained expensive for a very long time. There are, indeed, plenty of examples of people suffering today from the price of pharmaceuticals. The drift toward lower prices doesn't reach everywhere, everyone, or every drug at the same speed.

On the other hand, drugs and other methods that can keep us young won't just stop aging. They slow or lessen or stop the diseases related to aging, which when you consider the weakening of the immune system that comes with old age, is almost all of them.

Very few people would say that we shouldn't cure cancer or heart disease because only the wealthy will be able to afford it—and those who did would be seen by most as anti-human or insufferably whiny.

Seen in this light, it becomes obvious that this whole "only the rich will get hyperlongevity" mentality is pathetic in the extreme—a concession of defeat before the outset. If you think optimal health and longevity should be distributed, you won't say, "Well, it won't be distributed so I'm against it." You will try to make sure it gets distributed.

# MARTINE ROTHBLATT

*(See also Transbemanism, Transgender)*

In 1993, Martin Rothblatt became Martine Rothblatt. Having experienced the fluidity of genders under harsh conditions (i.e. surgery), Rothblatt became a leading transhuman advocate for the fluidity, diversity, and changeability of human, posthuman, and digitized life both for the individual and for the human species. She coined the phrase Transbemanism (see entry), the notion that humans should be identified by their thought patterns as opposed to their bodies. In 2006, Rothblatt started the Tesseract Movement, which describes its mission as "to promote the geoethical (world ethical) use of nanotechnology for human life extension," and to "conduct educational programs and support scientific research and development in the areas of cryogenics, biotechnology, and cyber consciousness."

An early enthusiast for the 1970s space colony concepts of Gerard K. O'Neill, Rothblatt earned a law degree and did work for various movers and shakers in the world of space exploration including the Hughes Space and Communications Group and NASA. In 1997, she led the group that formed the UN Declaration on the Human Genome and Human Rights that states, among other things, that "Genetic data associated with an identifiable person and stored or processed for the purposes of research or any other purpose must be held confidential in the conditions set by law."

Rothblatt's emphasis on joyful diversity contrasts with some of the more stodgy, or, one might say, mechanical transhumanist sorts, making her stand out when she speaks at transhumanist events.

**Roz Kaveney:** If anyone is going to persuade us to abandon the flesh, be downloaded, and live forever as information, it is Martine Rothblatt. A satellite scientist of distinction (she started GeoStar and Sirius Satellite Radio), who succeeded in saving the life of the child of her and her partner by throwing money at research, and saved scores of other children by doing so, she decided a few years ago that the next thing to conquer was death itself.

Martine believes that the philosophical opponents of uploading are mostly biological essentialists, people who believe that there is an absolute value in remaining true to an original biological form. The underlying logic of this position is that we are not clever enough to realize all the bad consequences of changing the naturally evolved order in any way. They think—and they have always thought—that we will come to regret any change to this.

# THE MATRIX

## (See also Transhumanist TV, Film, and Games)

We assume our readers already know about *The Matrix* (directed by the Wachowskis, 1999), the first of a trilogy that places humans in a simulated world generated by sentient machines that are, meanwhile, utilizing the humans as an energy source. In it, the hero Neo (Keanu Reaves) has an excellent adventure trying to escape The Matrix and defeat its overlords.

## THE CONSPIRACY

The Wachowski films connected with both popular culture and techno-intellectual theorists in a variety of ways. For some, it was about the quasi-political paranoia that we are all trapped by an unseen control system. Indeed, many conspiracy theorists have speculated that we may, in fact, be in a simulacrum under the control of a malevolent force of some sort (usually in cahoots with whoever is currently president of the United States). The first such "conspiracy nuts" may be the Gnostics, who go back to ancient Greece. They believed that the apparent material universe was under the control of a demiurge, a sort of false god who hides the "Logos," or the true reality, behind appearances.

## VIRTUAL REALITY: AN IDEA SO SCARY WE COULDN'T WAIT TO GET OUR HANDS ON IT

*The Matrix* was also a manifestation of the 1990s' intellectual and pop cultural fascination with virtual reality. Mind you, there wasn't much actual virtual reality happening at the time (or since). It was the *idea* of virtual reality that fascinated: being able to interact, via the saturation of (at least) our visual sense, in fully realized computerized worlds.

## THE TRENDY FRENCH PHILOSOPHER

The French postmodernist philosopher Jean Baudrillard, during this same time, was also popularizing his notion of The Simulacrum—the idea that we were living (even prior to virtual reality) in a condition of such media saturation that much of our public and political lives proceeded without any reference to anything real.

## IT COULD BE TRUE!

Finally, *The Matrix* appeals to those interested in Simulation Theory (see topic). We could be living, right now, inside a world created by a bratty and cruel twelve-year-old hacker from another dimension. That would explain a lot.

## THANK YOU FOR ACKNOWLEDGING US, WACHOWSKIS

*The Matrix* is generally well liked by the transhumanist crowd. Even though the films are, in part, trashy thrillers, they present enough of these high concepts to provide some frisson for those who hunger to have their interests intelligently reflected in popular film.

### Kurzweil Gives It One Thumb Up

**Ray Kurzweil (interviewed by R.U. Sirius and Surfdaddy Orca):** One problem with a lot of science fiction—and this is particularly true of movies—is they take one change, like the human-level cyborgs in the movie *AI,* and they put it in a world that is otherwise unchanged. There's no virtual reality, but you had human-level cyborgs.

I thought *The Matrix* was pretty good in its presentation of virtual reality. And they also had sort of AI-based people in that movie, so it did present a number of ideas. Some of the concepts were arbitrary as to how things work in the matrix, but it was pretty interesting.

## Hey! Cyberpunk Godfather William Gibson Had a Matrix, Too

**Surfdaddy Orca:** Before Neo and Morpheus in the seminal 1999 film, *The Matrix*, there was William Gibson. In Gibson's fiction, the matrix is the next generation of the Internet: an abstract three-dimensional grid, representing everything and everyone connected by arbitrary objects and avatars of their choosing.

Famously, Gibson characterizes this vision in the 1984 novel *Neuromancer*: "A graphic representation of data abstracted from banks of every computer in the human system. Unthinkable complexity. Lines of light ranged in the nonspace of the mind, clusters and constellations of data. Like city lights, receding."

Movement in Gibson's matrix occurs by thought, and feedback is directly given into the brain by electronic signals.

Early on—well before the Internet became as ubiquitous as the telephone—Gibson saw that critical decision-making involving massive amounts of complex multidimensional data requires more advanced and interactive visual representations than simple tables, pie charts, and line charts.

## Will "Trapped in The Matrix" Become a Diagnosis?

**Mikita Brottman:** Fashions in technical innovation have an important influence on the form, origin, and content of our psychological anxieties. As soon as a particular technology is widespread enough to be incorporated into a culture's mental life, it will be incorporated into that culture's paranoid or psychotic delusions. Psychiatric journals have recently contained reports of patients with delusional belief that they are characters in particular computer games, or that they are living inside reality television, or trapped in "The Matrix."

# MAX MORE AND NATASHA VITA-MORE

As noted previously, Max More and his wife Natasha Vita-More were the power couple that initiated contemporary transhumanism as a social movement. Max wrote one of the important initiatory documents of the movement with his "The Principles of Extropy." He lent his strong libertarian and actualist influences to the early days of the movement, but seems to have moderated those views somewhat since then.

Natasha is the artistic half of the couple, but has also written theoretical papers, including the very early (1983) "Transhumanist Manifesto." A bodybuilder and strong supporter of morphological freedom and body sculpting, Natasha's *Primo Posthuman* is a conceptual art piece built around radical body design. Natasha says she likes to think of it "as a cross between Frank Lloyd Wright, Le Corbusier, and Valentine."

They make a handsome couple.

**Ben Goertzel:** Way back in 1988, before the future was fashionable, Max More cofounded the original transhumanist magazine, *Extropy: The Journal of Transhumanist Thought.* In the early '90s he founded Extropy Institute, organized five Extro conferences—the first explicitly transhumanist conferences—and founded the English cryonics organization Alcor-UK (originally Mizar).

But his greatest achievements are perhaps on the intellectual rather than organizational side. Max has authored a host of seminal transhumanist essays, forming a key part of the foundation of modern transhumanist thought, including "Transhumanism: Toward a Futurist Philosophy" (1990), "The Principles of Extropy," "A Letter to Mother Nature: Amendments to the Human Constitution" (1999), "Technological Self-Transformation: Expanding Personal Extropy" (1993), "Dynamic Optimism: An Extropian Cognitive-Emotional Virtue" (1992), and more recent papers such as "True Transhumanism" (2009). He has also spoken at dozens of futurist conferences and spread transhumanist ideas through numerous newspaper, magazine, and TV interviews.

## Natasha Vita-More

**R.U. Sirius:** In Vita-More's 16mm film *Waking Goddess / Sleeping Muse,* she hiked thirty miles inside the Haleakală Volcano on the island of Maui and performed a soliloquy to the loss of her pregnancy. This artistic interpretation of women and the Earth led to Vita-More's performative works, which led her to the Amazon Jungle, Red Rocks Amphitheater, the US Space and Rocket Center's Space Camp training, and toward the future and eventually transhumanism.

# MEMORY-EDITING DRUGS

*(See also Cognitive Enhancement, Neurotechnology, Performance Enhancement)*

If extreme longevity is the premier obsession of transhumanists, getting the most out of everybody's second favorite organ—the brain—is possibly the second. Transhumanists want all possible brain enhancements, including increased memory. But sometimes forgetting—losing or softening traumatic memories—is a life enhancement . . . or is it?

The existential jury is still out about the value of consciously using a drug to edit memories of actual experiences, and I suppose there may never be a broad agreement as to whether it's a good or bad thing. On the other hand, cognitive science is lately telling us that many of the things that we think we remember are already distorted—or pre-edited—by our brains anyway, so what the hell. In fact, I'd like to erase every great movie I've ever seen, so I could view them anew. But for the moment, the neurological sciences are focused on traumatic memories.

**Surfdaddy Orca:** *The New York Times* recently reported on an announcement by Dr. Todd C. Sacktor and André A. Fenton at SUNY Downstate Medical Center, in which the doctors stated that a substance called PKMzeta is responsible for memory-related tasks in the brain. The *Times* characterized this as an "open door to editing memory." Studies with rats and mice show that a drug called ZIP interferes with PKMzeta, erasing learned behaviors. Editing memory may also mean enhancing the ability to remember things.

Fenton, who specializes in spatial memory in mice and rats, devised a way to imprint animals with memories for where things are located. He taught them to move around a small chamber to avoid a mild electric shock to their feet. Once rats learn, they do not forget. Placed back in the chamber, they remembered how to avoid the shock.

But when injected with ZIP (a zeta inhibitory peptide) directly into their brains, they had to start over again and learn how to avoid the electric shock. "When we first saw this happen, I had grad students throwing their hands up in the air, yelling," Dr. Fenton said. "Well, we needed a lot more than that one study."

Twelve other researchers have now independently confirmed Sacktor and Fenton's findings—essentially resolving a fundamental question of neuroscience about how memories are stored. A jointly published paper describes how long-term memories are stored as physical traces in the brain using a molecular mechanism of long-term memory storage.

This raises many questions. If human memory can be erased like a computer's hard drive, what happens to the "overwritten" memories? Is there a biochemical equivalent to disk restoration software? Can the erased memories ever be recovered? How does this affect learned behavior?

If memories can be edited—erased, enhanced, or supplanted—what replaces them? Several novels and short stories by science fiction master Philip K. Dick explore this idea. His novelette, *We Can Remember It for You Wholesale*, loosely adapted as the movie *Total Recall*, describes a man who has "extra-factual memory" implanted so that he remembers things he supposedly never did, like visit Mars.

Is editing a memory in some sense editing the self? Could I project an alternate identity—perhaps even multiple identities—and selectively edit memories to realize that identity or identities as me? This is a concept familiar to socially networked Internet users of virtual worlds such as World of Warcraft or Second Life. Alts—alternate identities—give users the ability to experiment with alternate versions of themselves.

The downside of using ZIP-like drugs to erase memory is spelled out by Nobel Peace Prize winner and Holocaust survivor Elie Wiesel:

> The authors and followers of the heinous 'Final Solution' were guilty not only of their unutterable crimes, but also of the will to erase their traces from the memory of others . . . Indeed they killed their victims two times: first with guns or in the gas chambers, and then by obliterating their memory. This is why I am somewhat hesitant to trust the proposed therapeutic means to use forgetting as a tool for healing. Once forgetting has begun, where and when should it stop?

# THE METHUSELARITY

*(See also Aubrey de Grey, Longevity/Immortality,*
*SENS Research Foundation)*

Will there be a tipping point for longevity? The Methuselarity, playing on the concept of The Singularity, is the idea that there will be a point in time when life extension technology slows aging down to zero. In other words, when the healthspan of the body is increasing by more than a year, every year, that's The Methuselarity (named for the oldest man in the Bible, Methuselah). From that point on, something like immortality becomes plausible, or so your friendly neighborhood immortalist would say. In the words of Aubrey de Grey, "The Methuselarity is a name my friend Paul Hynek has given to the point at which we reach what I have called longevity escape velocity. Longevity escape velocity (LEV), in turn, is the rate at which therapies to repair the molecular and cellular damage of aging need to be improved in order to stop their recipients from becoming biologically older."

# MIND-READING BOTS

*(See also Neurotechnology)*

On April 22, 2010, a short cryptic statement appeared in Japan's largest business newspaper, *Nikkei Asian Review,* announcing a goal to make available commercial mind-reading devices and personal assistant bots within the next ten years.

It's not surprising that the Japanese government and private sector would collaborate on a new initiative to develop bots with AI capable of detecting when you're hungry, cold, or in need of assistance, and electronics that can be controlled by thought alone. BMI (Brain-Machine Interface) technology typically involves an EEG sensor connected to a computer that can be controlled purely by thought (or, more accurately, brainwaves). Research and early prototypes include full helmets, headbands, and direct brain implants to capture and interpret brainwaves.

While the US Army actively pursues "thought helmets" that might someday lead to secure mind-to-mind communication between soldiers, the Japanese are going after the consumer market. The aim is to produce BMI technology to change TV channels or to use mobile phones to send text messages composed by thought alone. Hitachi is pursuing thought-controlled TV, promising something in years, not decades.

The *Nikkei* article suggested that Honda and Toyota will also be involved in the Japanese brainwave initiative. This might include robots that know when an elderly or physically disabled person needs help carrying a heavy load. The Honda ASIMO robot is already billed as the world's most advanced humanoid commercial robot. Toyota already has a thought-controlled wheelchair with a stylish EEG cap system, which has already created quite a stir.

## WHAT'S THE BIG IDEA?

The idea behind BMI is to control brainwaves and detected brain blood-flow patterns through ever more sophisticated and commercially available sensor-mounted headsets. Other applications mentioned in the *Nikkei* announcement include a car navigation system that searches for restaurants when the driver thinks of having a meal, and air-conditioners that adjust the temperature when people in the room feel too warm or cold.

## NETWORKED BOTS

The ATR Intelligent Robotics and Communication (IRC) lab is particularly interesting in its research focus. The lab makes cool stuff that portends a future of networked Japanese social robots assisting old people in their thought-controlled Toyota wheelchairs (let's hope the brake pedals work properly!). The lab is working on a measurement tool to easily locate people in the vicinity of a social robot in congested public spaces. When approaching a person, the robot first recognizes your face when you stand in front it. With a BMI interface, it will not be necessary to make a verbal request to the robot. The robot will guide you to the place you want to go.

A networked robot system can better provide information and guide people by coordinating a team of robots as well as Internet agents with other embedded devices such as cameras, electronic tags, and wearable sensors. The IRC lab envisions coordination between three different kinds of bots: "visible" (think ASIMO), "virtual" (think avatar), and "unconscious" (think embedded sensor), all in a cooperating system to provide a complete set of social services to both augmented and non-augmented humans in the urban environment. Such a networked robotic system would possess the intelligence to modify its communication techniques—including thought, speech, and gesture—to meet the needs of the current situation.

Ultimately, whether the BMI is a tiny wireless chip that has been surgically implanted or a less invasive EEG headband, it's likely that the simple thought of logging on to the Internet will soon link you to your fellow humans and intelligent bots. Reading between the lines of the recent *Nikkei* announcement—and understanding the cultural milieu and social nature of robotics in Japan—it's easy to see how Japan could easily emerge as a leader in BMI consumer electronics in the early 2020s.

*Written with Surfdaddy Orca*

# MIND UPLOADING

The idea of moving or copying a mind to a computer (sometimes called Whole Brain Emulation, or WBE) first appeared in science fiction in the 1950s. It's one of transhumanism's most fascinating and controversial topics. Ray Kurzweil is probably its most visible proponent, and he believes it will be possible by 2045.

## DIGITAL IMMORTALITY AND SPACE TRAVEL

If you could upload your mind, you'd be approximately immortal, both because Digital You would not be susceptible to aging like old-fashioned Meat You, and because you could make backups. Digital You would be much smarter, with a brain millions of times faster. Digital You would also be better suited for things like space travel, because moving, supplying, and protecting a computer in space is far easier than doing the same for a human body.

## TYPES OF MIND UPLOADING

This could happen in several ways. In what's called copy-and-transfer, your mind would be scanned and then a digital replica made. This might or might not involve destroying the original. Alternatively, parts of your brain might be replaced with prosthetic parts that worked in the same way, or better.

But why just be a brain in a box? Use a human or humanoid body, assuming you can obtain one nobody else is using. Maybe grow a clone, or install Digital You in a robot and become a Transformer.

## SO HOW CLOSE IS THIS?

Not close. Progress has been made, but there's a long way to go. The neural systems of roundworms, fruit flies, and even rodents have been simulated, at least to some degree. Computers grow ever more powerful, but a huge amount of raw computing power is needed to emulate the human mind. Quantum computers might be especially useful, in part because some think that consciousness relies on quantum effects. If Moore's Law continues to hold, we might have the necessary hardware in a few decades.

Of course, even the most powerful computer won't help if you don't deeply understand what you are simulating, and we still don't know how the mind really works. Learning, memory, consciousness, and much else are still only partly understood. If we can't explain how such things happen in the world of neurons, synapses, neurotransmitters, hormones, and all the rest, we can't hope to make it all work artificially.

Most psychologists would probably scoff at the thought of copying, transferring, or emulating something as complex as the human mind without screwing it up in some way, because they know how similar healthy and unhealthy minds can be. However, neuroscience, brain imaging, biotechnology, and related fields continue to progress, so the related practical issues may be solved by the time the raw computing power arrives.


## PHILOSOPHICAL ISSUES

Transhumanism often brings up questions of ethics, but mind uploading brings up more purely philosophical issues than any other topic. For example: Is consciousness something that can be separated from the brain? If the mind is essentially separate from its biological substrate, it shouldn't be impossible to move it to a different substrate. On the other hand, consciousness might be an emergent property of biology, and not reproducible on a different substrate. Even a computer as powerful as a human brain might still lack crucial aspects needed for consciousness.

Let's assume your mind is perfectly copied to a computer. In what sense is this new Digital You really "you"? Digital You may be conscious, think like you, and have all your memories, but plain old Meat You is still here. If Meat You dies, an outside observer might say "you" live on as Digital You, but the inside observer who used to be in Meat You is still gone.

There are plenty of ethical, legal, and comic possibilities. In a practical sense, who or what is this new being, Digital You? Does Digital You count as "alive"? Does Digital You have rights? Would destroying him be murder? Can he legally drive your car and use your credit card? Can you put Digital You to work and keep the paycheck? Is he responsible for a share of the mortgage? What about your wife? Is Digital You also married to her? Can she divorce one without divorcing the other? Can Digital You vote? If you made a hundred copies, could they *all* vote? If Meat You dies, who gets your stuff? Clearly, The Singularity will not eliminate the need for lawyers.

## Five Main Developments Leading to Mind Uploading

**Randal Koene and Ben Goertzel:** There are five main developments in the last decade that I would consider very significant in terms of making substrate-independent minds a feasible project.

The first is advances in computing hardware, first in terms of memory and processor speeds, and now increasingly in terms of parallel computing capabilities. Parallel computation is a natural fit to neural computation. As such, it is essential both for the acquisition and analysis of data from the brain, as well as for the re-implementation of functions of mind.

The second major development is advances in large-scale neuroinformatics. That is, computational neuroscience with a focus on increasing levels of modeling detail and increasing scale of modeled structures and networks.

Third, actual recording from the brain is finally beginning to address the dual problems of scale and high resolution. A much-celebrated radical development is optogenetics, the ability to introduce new channels into the synapses of specific cell types, so that those neurons can be excited or inhibited by different wavelengths of light stimulation. It is technology such as this, which combines electro-optical technology and biological innovation, that looks likely to make similar inroads on the large-scale, high-resolution neural recording side.

A fourth important type of development is projects that are aimed at accomplishing whole brain emulation. There we see tool development projects, including three prototype versions of the Automated Lathe Ultramicrotome (ATLUM) developed at the Lichtman Lab at Harvard University, which exists to acquire the full neuronanatomy, the complete connectome of individual brains for reconstruction into a whole brain emulation.

The fifth development during the last decade is somewhat different, but very important. It is a conceptual shift in thinking about substrate-independent minds, whole brain emulation, and mind uploading. Ten years ago, I could not have visited leading mainstream researchers in neuroscience, neural engineering, computer science, nanotechnology, and related fields to discuss projects in brain emulation. It was beyond the scope of reasonable scientific endeavor, the domain of science fiction. This is no longer true.

## The Two Questions: Feasibility and Identity

**Extropia DaSilva:** There are two major questions surrounding the concept of mind uploading. There is the question of feasibility: Can we build a model of a brain complete enough to allow a conscious mind to emerge? The other question is concerned with identity. Some people argue that, if a copy of a conscious mind is identical by all measures (ignoring the fact that one is biological and the other is neuromorphic software/hardware) it should be thought of as a continuation of the mind that was mapped and uploaded. Others argue that a copy cannot be considered the same as the original, so the newly awakened consciousness must be another person.

Various attempts have been made to imagine the benefits of mind uploading. Assuming continuation of the mind, these benefits include indefinite lifespans and upgrading the mind. When your current brain no longer works well enough—or at all—you transfer your conscious mind to another (perhaps better) artificial brain. None of these benefits are tempting to those who see uploads as different people.

Hans Moravec said, "I consider these future machines to be our progeny, 'mind children' built in our image and likeness." I think "mind children" is an appropriate term for uploads. After all, when you have offspring some of your genes are duplicated. Each person is the result of two channels of heredity: genetic data encoded in DNA, and culture. Vernor Vinge has suggested humans can be defined as the species that learned to outsource aspects of cognition. This began with the evolution of complex language and the ability to communicate thoughts and intentions to other minds. It continued with the emergence of writing and the preservation of memories in external systems, and now includes technologies that are gradually assuming functions once thought to be exclusively biological. Mind uploading would mark the point at which culture becomes completely independent of biology—when all (not just aspects of) a person's cognition could be duplicated. Can it really be the case that people would treat their upload as a stranger? I would imagine there would actually be a connection that is closer than that which can exist between identical twins.

## Getting Past Fixed Identity in Preparation for Uploading

**Extropia DaSilva:** Traditionally (in the West at least), the self has been attributed to an incorporeal soul, making "I" a fixed essence of identity. But neuroscience is revealing the self as an interplay of cells and chemical processes occurring in the brain—in other words, a transitory dynamic phenomena arising from certain physical processes. There isn't any self in that sense; rather (in author and science journalist Lone Frank's words):

> Life is not so much about finding yourself but choosing yourself or molding yourself into the shape you want to be . . . The neurotechnology of the future will likewise produce the means for transforming the physical self—be it through various cognitive techniques, targeted drugs, or electronic implants . . . our individual self will simply be a broad range of possible selves.

By the time mind uploading is generally available, perhaps people will have forgotten a time when a singular self was "normal." They will be used to multiple viewpoints, their brains processing information coming not only from their local surroundings, but also from the remote sensors and cyberspaces they are simultaneously linked to. They will have already become familiar with mental concepts migrating from the brain to spawn digital intermediaries within the clouds of smart dust that surround them. Every idea, each inspiration, would give birth to software lifeforms introspecting from many different perspectives before integrating the results of their considerations with the primary consciousness that spawned them.

## Know Your Uploaded Rights

**Roz Kaveney:** Before we start uploading ourselves, people need to possess intellectual property in themselves. The uploaded need legal personhood.

## Don't Get Hacked

**Roz Kaveney:** It is easy to brainwash the embodied, and the uploaded would likely be even more vulnerable. We have to establish an ethic of the absolute impermissibility of harming a person's autonomy by harming their own value to themselves. It is an assault. Martine Rothblatt: "Minds are fragile—and to hit someone in the face is almost better than to put a fist through the fragile web of a personality."

## Skeptic

**Charles Stross (interviewed by R.U. Sirius and Paul McEnery):** Mind uploading would be a fine thing, but I'm not convinced what you'd get at the end of it would be even remotely human.

# MOORE'S LAW

*(See also Law of Accelerating Returns)*

In 1965, Intel co-founder Gordon Moore observed that the number of transistors on integrated circuits doubles approximately every two years. Since that time, this "law" has essentially remained unbroken. In fact, the computer hardware industry has adopted the law as a tautological goal, making it arguably the only industry in which making progress shares equal importance with making money.

Every few years, word goes out that we will break Moore's Law and the great dreams of techno-optimists will be handcuffed, but then something comes along to help us shrink the circuits again.

Recently the talk is that the circuit has been just about shrunken to its usable limits, but even if that's the case, the increase of processing power may continue unabated with the help of such technologies as graphene and quantum computing (see entries).

Of all the transhuman hopes and dreams, The Singularity is most dependent on the continued "enforcement" of Moore's Law.

## What Happens When Computer Components
## Just Can't Get Any Smaller?

**Warren Frey:** For all the optimism about humanity's impending ascent into the digital realm, writ large with logarithmic graphs from Ray Kurzweil and given life by the fiction of Charles Stross, there's an obstacle we haven't been able to bound over. Inevitably, we reach the point where computer components just can't get any smaller and still work in the realm of the electron. Moore's Law, the phenomenon of computers doubling in power while plunging in price, has to end, unless there are new developments that push us into new terrain. One such development may allow future electronics to shed the electron and embrace light, not only as the resource behind ever faster and denser digital communications, but as a way to look at the world.

In 2009, at the NSF Nanoscale Science and Engineering Center in Berkeley, California, Dr. Thomas Zentgraf and his colleagues achieved a major breakthrough. They were able to guide light at a nano scale. Zentgraf told me, "To generate light at the nano scale, you put a light source on a chip, then combine it with optics so you can generate and guide light around. It isn't on a computer chip at the moment, but I'm personally optimistic we'll see chips like this in ten to twenty years."

With these chips, we can hopefully have a new path when traditional electronics runs up against its limits, and Moore's Law starts to look more like a temporary statute. Electrons simply can't go much further without running into the laws of physics. That's where light has distinct advantages.

# MORMON TRANSHUMANIST ASSOCIATION

What could better express the delightfully strange diversity of the transhuman world than the existence of the Mormon Transhumanist Association (MTA)? Just learning of its existence usually raises titters among transhumanist types. Most transhumanists are devout atheists, albeit some have argued that their faith in technological progress is more or less akin to a religion. But the MTA persists in being nice around their fellow transhumanists. This, and other dastardly deist tricks, makes the MTA a hard-to-ignore part of the overall transhumanist scene.

# The Mormon Transhumanist Speaks

**Lincoln Cannon (interviewed by Hank Pellissier):** Most Mormon Transhumanists consider our religion to be remarkably compatible with transhumanism. We consider Mormonism to be a religious transhumanism. Eternal progression is a central doctrine of Mormonism. Basically, the idea is that we have all existed in some form or another into the indefinite past; that we have been and are progressing toward becoming like God in a creative and benevolent capacity; and that we should each help others do the same into the indefinite future. Mormon scripture asserts the work of God to be that of bringing about immortality and eternal life, and invites us all to participate in that work.

Mormon scripture also situates us in the "Dispensation of the Fullness of Times," when God is accelerating the work, inspiring us with greater knowledge, and endowing us with greater power in preparation for the prophesied millennium, a time of transfiguration, immortality, resurrection, renewal of this world, and ultimately the discovery and creation of worlds without end. Early Mormon prophets, Joseph Smith and Brigham Young, suggested that we would begin performing the ordinance of transfiguration before the millennium, and that immortals would begin performing the ordinance of resurrection during the millennium.

# Mormon Cosmology Is Far Out!

**Lincoln Cannon (interviewed by Hank Pellissier):** Mormon cosmology, as articulated in Mormon scripture, includes the idea that God, in whom we should all participate, creates worlds without end, heavens and glories without end, each according to the desires of its inhabitants, according to that which they are willing to receive. While I do not subscribe to mere moral relativism, I do value this idea of an indefinitely broad and deep cosmos, organized and reorganized in a perpetual work to fulfill desires, wills and laws, overcome conflicts and tensions, and provide time and space enough to explore ourselves and each other, and experience a full measure of our creative capacities. Something like that, in my estimation, is godhood, so far as I can imagine it.

# NANOTECHNOLOGY

*(See also Foresight Institute, Graphene, NBIC)*

Nanotechnology is the manipulation of matter on the scale of atoms and molecules. Its impressive achievements and mind-boggling possibilities put it at the core of transhumanist hopes. The term covers everything from particles (now used in cosmetics, sunscreens, medicine, clothing, and even condoms), to advanced macroscale materials (like graphene), and the still largely hypothetical field of molecular nanotechnology (nanomachines and nano-assemblers).

## SMALLER, STRONGER, LIGHTER, BETTER

Nanomaterials can behave in unusual ways, often due to their high surface area to mass ratio. In addition, precisely arranging atoms and molecules allows the creation of novel materials that are stronger, lighter, more efficient, greener, and better in many ways. Nanotech is a revolutionary, general-purpose technology like steam engines and electricity. It's revolutionizing many industries, with too many developments to list and more happening all the time. If it advances as expected by its advocates, it will change the 21st century, and maybe help bring on The Singularity.

## A NANOHISTORY

Richard Feynman's 1959 talk "There's Plenty of Room at the Bottom" is often considered the starting point, but the field didn't really hit scientific and public consciousness until the 1980s. That's when the scanning tunneling microscope and the atomic force microscope were invented, "buckyballs" and other fullerenes were discovered, and K. Eric

Drexler published *Engines of Creation: The Coming Era of Nanotechnology* and cofounded The Foresight Institute. Nanotech got going and hasn't stopped. There are already thousands of nanotech products, and every year, billions of dollars are invested in nanotech research and development.

## TOP-DOWN VS. BOTTOM-UP

There are two basic ways to construct nanotech materials and devices. Top-down means using larger machines to create smaller ones. It's what microchip manufacturers do with photolithography. Techniques like atomic layer deposition, atomic force microscopes, and focused ion beams can also be used for nanoscale fabrication. Bottom-up approaches include standard methods of chemical synthesis, as well as more advanced methods of nudging molecules to self-arrange into a desired shape.

Mechanosynthesis, a more precise version of conventional chemosynthesis, is a major goal of molecular engineers. Ribosomes do it naturally, in a bottom-up way, but so far researchers can only do it top-down, with great difficulty, using tools such as atomic force microscopes. However, you don't want to build things made of millions of atoms by manually placing one atom at a time. For that you need automation, or (better yet) some form of bottom-up self-assembly. If and when that happens, nanotech will remake the world.

## NANOTECH TODAY

Already available or in development are stronger steel, plastics, ceramics, composites, and ultra-high-performance materials of all sorts. Nanotech coatings enable glass to clean itself and textiles to repel water and stains. Nanotech is improving solar cells, batteries, fuel cells, computer chips, and flat-panel displays. Better catalysts are improving chemical production, and nanotech filters are removing arsenic, chemicals, metals, viruses, and even salt from water.

## NANOMEDICINE

Nanomedicine is already a multi-billion-dollar industry, mostly due to improved drugs and drug delivery methods, and is also allowing better biomedical sensors and previously impossible means to precisely intervene in the body.

In an outgrowth of their work with silicon wafers, IBM is now making nontoxic, biodegradable, nanoscale "ninja polymers" that act as antibiotics, but in a new way. These polymers are drawn to and physically pierce the cell walls of antibiotic-resistant superbugs like MRSA. That makes it much harder for the targeted organism to develop resistance.

Liangfang Zhang, a professor of nanoengineering at the University of California, San Diego, has developed "nanosponges" that could become the basis for an entirely new type of vaccine. Membranes from red blood cells are wrapped around nanoparticles to become decoys, so small that one red blood cell can supply enough material for thousands. Toxic proteins secreted by dangerous bacteria get soaked up, trapped, and are cleared from the body. Without its toxins, the bacteria is also more vulnerable to the immune system.

Researchers at North Carolina State University and the University of North Carolina at Chapel Hill have developed nanoparticles that deliver two different cancer-killing drugs to separate parts of cancer cells. An outer shell tricks receptors on cell membranes into grabbing the nanoparticle and breaking down the shell, which releases a drug that destroys the cell membrane. The breakdown of the shell then releases a second drug that attacks the nucleus of the cancer cell. This one-two punch makes it harder for the cancer cells to develop resistance.

UCLA researchers have discovered that nanodiamonds can treat some forms of cancer, and could be used to promote bone growth and improve dental implants.

## FARTHER OUT

Hopes for longevity treatments often rest on nanotech. The dream is to use programmed nanomachines (a.k.a. nanorobots, nanobots, or nanites) to target and destroy cancer cells, viruses, and plaques in the arteries and the brain. Imagine curing cancer, AIDS, coronary heart disease, and Alzheimer's with one injection. Nanotech might also help us transition to a solar energy based economy, prevent fresh water shortages and famines, reduce resource depletion, and solve global warming by removing excess $CO_2$ from the atmosphere.

Dr. J. Storrs Hall has used the term "utility fog" to describe a swarm of networked nanobots that can reconfigure itself. Connecting their tiny arms as needed, a bicycle could become a backpack for carrying in the elevator, and then in your office become your personalized chair. Also, things at the atomic level move fast, meaning that a *Star Trek* replicator could fabricate your entrée from atoms faster than you could cook it.

For a good fictional treatment of a world transformed by nanotech, check out Neal Stephenson's 1995 novel, *The Diamond Age*. (The strength and many uses of carbon bonds make it a good material for building nanomachines, so it's expected that many will be built out of what is essentially diamond.)

## CHALLENGES AND FEARS

Despite continuing progress, we still don't have nanobots. At that scale, things like viscosity and Brownian motion are major hurdles. In the opinions of critics, these may be insurmountable obstacles.

There are valid, near-term concerns about nanotech health hazards. Antimicrobial nanosilver in clothing washes out in the laundry and enters the environment, to unknown effect. Nanoscale particles can easily penetrate intact skin, or enter the body through the lungs or the gastrointestinal tract. Carbon nanotubes are physically similar to asbestos fibers, and some studies indicate they may be similarly hazardous.

A self-replicating machine was made of Legos in 2002, demonstrating (in a very limited way) the classic nanotech fear of runaway replicators turning Earth into "grey goo." Thankfully, responsible researchers avoid making anything self-replicating that might go out of control, and it's much harder to build a replicator that can cope with the natural environment.

Like even the best technology, nanotech promises to bring economic disruptions along with benefits, and people rightly fear the use of nanotech as a weapon. But whatever happens, it's very likely that advanced, atomically precise nanotechnologies will transform the 21st century.

**Ben Goertzel:** Contemporary nanotech mostly focuses on narrower nano-engineering than what Eric Drexler envisioned, but arguably it's building tools and understanding that will ultimately be useful for realizing Feynman's and Drexler's vision.

**J. Storrs Hall (interviewed by Surfdaddy Orca):** My best guess is that the technology to win the Feynman Grand Prize (that is, build a working robot arm at the nanoscale) will come somewhere between 2020 to 2030, depending on the level of effort.

**Ralph Merkle (interviewed by Surfdaddy Orca):** We can do things today at the molecular scale that we couldn't do even a few years ago. Still, it might be a few decades before we have the ability to inexpensively arrange atoms in most of the ways permitted by physical law, but progress has been remarkable—the exponential trends continue to be exponential, and there's every reason to think they will continue that way for some time.

# DNA Origami

**Surfdaddy Orca:** A promising molecular-level manufacturing technology that can also be used for targeted drug delivery is found in Paul Rothemund's work folding stringy DNA molecules into tiny, two-dimensional patterns known as DNA origami. Like diamondoid mechanosynthesis, DNA origami is a bottom-up fabrication technique that exploits the intrinsic properties of atoms and molecules to make simple nanostructures.

## Self-Assembling Nanoparticles into Complex Nanostructures

**Surfdaddy Orca:** Recently, a team led by Dr. Ting Xu at the US Department of Energy's Lawrence Berkeley National Laboratory made an important nanotech advance. They found a simple and yet powerful way to induce nanoparticles to assemble themselves into complex arrays. By adding specific types of small molecules to mixtures of nanoparticles and polymers, Dr. Xu's group directed the self-assembly of the nanoparticles into arrays of one, two, and three dimensions with no chemical modification of either the nanoparticles or the block copolymers. In addition, they found that the application of external stimuli—light and heat—can be used to further direct the assemblies of nanoparticles for even finer and more complex structural details.

The nanofactories of tomorrow will likely require both molecular manufacturing as envisioned by Eric Drexler using chemical synthesis and nanomanufacturing techniques like Dr. Xu's. Nanoparticles can now be induced to self-assemble nonchemically using block copolymers in regular patterns on surfaces and at interfaces to provide better data storage, solar cells, and tiny electronics.

# NBIC

There's no one silver bullet technology that's going to make us all into superbeings. The trick is in the mix of different technologies. NBIC stands for nanotechnology, biotechnology, information technology, and cognitive science.

Look at it this way: with nanotechnology and biotechnology, we stand to gain control over inorganic and organic matter. With advanced information technology, we get not just the horrendous data glut that's now tormenting us (and our National Security Agency), but data that's shaped, analyzed, and made useful by productive machines that can then act on the data. With cognitive science, we get the maximum use of that thing inside our heads that is allowing me to produce and you to consume these half-baked thoughts.

## YOU CAN'T HAVE ONE WITHOUT THE OTHER

These all merge together. We'll be using nanotechnology and neurotechnology to create information technology AGIs. We'll be using information technologies to understand our biological and cognitive data, and to make sure our nanotech molecular machines don't run amok and flood the world with Spongebob dolls.

Here's a brief summary of the NBIC technologies:

- **Nanotechnology:** Technology related to features of nanometer scale (ten meters), including thin films, fine particles, chemical synthesis, advanced microlithography, and so forth.

- **Biotechnology:** The application of science and engineering to the direct or indirect use of living organisms, or parts or products of living organisms, in their natural or modified forms.

- **Information technology:** Applied computer systems, both hardware and software, including networking and telecommunications.

- **Cognitive science:** The study of intelligence and intelligent systems, particularly intelligent behavior as computation.

The National Science Foundation (NSF) and a formidable-sounding government subcommittee called the National Science and Technology Council on Nanoscale Science, Engineering, and Technology have published a number of reports exploring the convergence of the NBIC technologies as the result of a series of conferences between 2001 and 2006. The chief application areas they've identified include:

- Expanding human cognition and communication,

- Improving human health and physical capabilities,

- Enhancing group and societal outcomes,

- Strengthening national security, and

- Unifying science and education.

The convergence, these reports suggest, will be based on the "unity of nature at the nanoscale," along with technology integration at the nanoscale, key transforming tools, and the pursuit of improvements in human performance. "A revolution is occurring in science and technology, based on the recently developed ability to measure, manipulate, and organize matter on the nanoscale—one to one hundred billionths of a meter," writes William Sims Bainbridge, co-director of Human-Centered Computing at the NSF and co-editor with Mihail Roco of several NSF publications on NBIC. "At the nanoscale, physics, chemistry, biology, materials science, and engineering converge toward the same principles and tools. As a result, progress in nanoscience will have very far-reaching impact."

"The human mind can be significantly enhanced through technologically augmented cognition, perception, and communication," writes Bainbridge. "Research will focus both on the brain and the ambient socio-cultural milieu, which both shapes and is shaped by individual thought and behavior." Specific technology includes personal sensory device interfaces and enhanced tools for creativity along with continued humanization of computers, robots, and information systems.

"Nano-bio sensors and processors will contribute greatly to research and to development of treatments, including those resulting from bioinformatics, genomics, and proteomics," suggests Bainbridge. Specific technologies include implants based on nanotechnology and regenerative biosystems that will start to replace human organs, and nanoscale machines unobtrusively providing needed medical intervention.

NBIC will likely be used to enhance intelligence, mobility, cognitive qualities, vision, and hearing. "I think we will stop short of eugenics but proceed to offer neurological and physical enhancements that improve the quality of life under the umbrella of medicine," writes James Canton of the Institute for Global Futures.

The following timeline, written in 2006, for converging NBIC technologies was tabulated from dozens of authors and presenters who participated in the first three NSF-sponsored NBIC conferences. Today, in 2014, we're too close to 2015 to believe in it, are we not? Yes, the official futurists seem exceptionally overoptimistic, proving the R.U. Sirius maxim that everything takes twice as long as you expect it to after you've already taken that into account. Or perhaps some government committee is overselling tech miracles to keep us sedate?

## 2015

- Anywhere in the world, an individual will have instantaneous access to needed information, whether practical or scientific in nature, in a form tailored for most effective use by the particular individual.

- Comfortable, wearable sensors and computers will enhance every person's awareness of his or her health condition, environment, chemical pollutants, potential hazards, and information of interest about local businesses, natural resources, and the like.

## 2020

- National security will be greatly strengthened by lightweight, information-rich war-fighting systems, capable uninhabited combat vehicles, adaptable smart materials, invulnerable data networks, superior intelligence-gathering systems, and effective measures against biological, chemical, radiological, and nuclear attacks.

## 2025

- Robots and software agents will be far more useful for human beings, because they will operate on principles compatible with human goals, awareness, and personality.

- The human body will be more durable, healthier, more energetic, easier to repair, and more resistant to many kinds of stress, biological threats, and aging processes.

- A combination of technologies and treatments will compensate for many physical and mental disabilities and will eradicate altogether some handicaps that have plagued the lives of millions of people.

## 2030

- Fast, broadband interfaces between the human brain and machines will transform work in factories, control automobiles, ensure military superiority, and enable new sports, art forms, and modes of interaction between people.

- The ability to control the genetics of humans, animals, and agricultural plants will greatly benefit human welfare; widespread consensus about ethical, legal, and moral issues will be built in the process.

## 2050

- The vast promise of outer space will finally be realized by means of efficient launch vehicles, robotic construction of extraterrestrial bases, and profitable exploitation of the resources of the Moon, Mars, or near-Earth asteroids.

*Written with Surfdaddy Orca*

# NEUROBOTICS

*(See also Robotics)*

According to Urban Dictionary, neurobotics is "the occupational field of building, studying, or designing robotics (or robots) that emulate 'pathetic humans.'"

While I doubt that those working in the field have picked out particularly pathetic humans to emulate (and who are they to judge?), Steven Kotler, upon visiting with Jeff Krichmar, the man who first described the neurobotics approach to robotics and AI back in 1978, concluded that the goal of neurobotics is to "build a brain from the ground up, one neuron at a time." Still others define neurobotics as building connections between neural and robotic parts that mimic the functioning of various bodily appendages.

**Steven Kotler:** Neurobotic science sits at the convergence of robotics, artificial intelligence, computer science, neuroscience, cognitive psychology, physiology, mathematics, and several different engineering disciplines. Computationally demanding and requiring a long view and a macroscopic perspective (qualities not often found in our world of impatient specialization), the field is so fundamentally challenging that there are only around five labs pursuing it worldwide.

## Bottoms Up!

**Steven Kotler:** Neurobotics is an outgrowth of a growing realization that, when it comes to understanding the brain, neither computer simulations nor top-down robotic models are getting anywhere close. As Dartmouth neuroscientist and Director of the Brain Engineering Lab Richard Granger puts it, "The history of top-down-only approaches is a spectacular failure. We learned a ton, but mainly we learned these approaches don't work."

Gerald Edelman, a Nobel Prize-winning neuroscientist and Chairman of Neurobiology at Scripps Research Institute, first described the neurobotics approach back in 1978. In his "Theory of Neuronal Group Selection," Edelman essentially argued that any individual's nervous system employs a selection system similar to natural selection, though operating with a different mechanism. "It's obvious that the brain is a huge population of individual neurons," says UC Irvine neuroscientist Jeff Krichmar. "Neuronal Group Selection meant if we could apply population models to neuroscience, we could examine things at a systems' level." This systems approach became the architectural blueprint for moving neurobotics forward.

## The Edge of Real Brain Complexity

**Steven Kotler:** The robots in Jeff Krichmar's lab don't look like much. CARL-1, his latest model, is a squat white trash can contraption with a couple of shopping cart wheels bolted to its side, a video camera wired to the lid, and a couple of bunny ears taped on for good measure. But open up that lid and you'll find something remarkable—the beginnings of a precise replica of a biological nervous system. CARL-1 has thousands of neurons and millions of synapses that, Krichmar says, "are just about the edge of the amount of size and complexity found in real brains." Not surprisingly, robots built this way—using the same operating principles as our nervous system—are called neurobots.

Krichmar emphasizes that these artificial nervous systems are based upon neurobiological principles rather than computer models of how intelligence works. The first of those principles, as he describes it, is: "The brain is embodied in the body and the body is embedded in the environment—so we build brains and then we put these brains in bodies and then we let these bodies loose in an environment to see what happens." This has become something of a foundational principle—and the great and complex challenge—of neurobotics.

When you embed a brain in a body, you get behavior not often found in other robots. Brain bots don't work like AIBOs (robotic pet toys). You can buy a thousand different AIBOs and they all behave the same. But brain bots, like real brains, learn through trial and error, and that changes things. "Put a couple of my robots inside a maze," says Krichmar, "let them run it a few times, and what each of those robots learns will be different. Those differences are magnified into behavior pretty quickly." When psychologists define personality, it's along the lines of "idiosyncratic behavior that's predictive of future behavior." What Krichmar is saying is that his brain bots are developing personalities—and they're doing it pretty quickly.

# NEUROTECHNOLOGY

*(See also Artificial Hippocampus, Cognitive Enhancement, Implants, Mind-Reading Bots, Mind Uploading, Optogenetics, Prosthetics)*

Neurotechnology is any technology that aids the study, repair, control, or enhancement of the brain. Most research has focused on studying the brain and on overcoming disabilities, but some of the most exciting work involves controlling computers and other devices, using the brain to directly communicate, and enhancing its function.

## NEURAL INTERFACES

There are many ways of getting data out of the brain (and sometimes sending it in, a.k.a. neuromodulation or mind control). All these methods can be used to control computers, though most predate them, and are considered brain-computer interfaces (BCIs).

Noninvasive methods include electroencephalogram (EEG), which uses scalp sensors to detect brainwaves. Invented in 1924, it's the oldest neural interface. Electromyogram (EMG) senses the electrical signals produced by muscle contraction and relaxation. Electrooculogram (EOG) reads the electrical signals produced by eye movements. More capable methods using more expensive neuroimaging technologies include functional magnetic resonance imaging (fMRI), positron emission tomography (PET), and magnetoencephalography (MEG).

Noninvasive methods are relatively risk-free and sometimes inexpensive. However, even the best noninvasive sensors are limited in resolution. Invasive methods use electrodes implanted in the skin, under the skull, or into the gray matter to get signals directly from neurons. These give higher resolution signals and (sometimes) two-way data transmission ability, but require surgery and risk scar tissue build-up. There's also optogenetics, which uses DNA and fiber optics to precisely target specific neuron types.

BCIs tend to be bulky and need to be tethered, but researchers at Brown University have created a low-power, wireless BCI that can be read remotely and recharged via inductive charging. Perhaps the leading edge of noninvasive sensors are the temporary, wireless "electronic tattoos" developed by Todd Coleman at UC San Diego. Electronic tattoos are thin, barely visible layers of plastic with embedded circuitry that senses brain waves or subvocalizations.

## USING THAT CONNECTION

Deep brain stimulation (DBS) uses an implanted "brain pacemaker" to treat Parkinson's, depression, or chronic pain. Surgeons at Johns Hopkins in Baltimore are testing the technique on patients in the early stages of Alzheimer's disease.

Neuroprosthetics are a major type of neurotechnology. Early versions were invented by Professor José Delgado in the 1960s, who used a radio-controlled "stimoceiver" to control many emotions in his animal and human subjects. Paranoids everywhere were thrilled to see proof of what they always knew.

Today's neuroprosthetics typically connect a part of the nervous system (not necessarily the brain) to a device. Examples include cochlear and retinal implants, but one of the greatest areas of excitement is the increasing ability of humans (not to mention monkeys) to control prosthetic body attachments with the brain. A 2008 experiment at the University of Pittsburgh gave a monkey a prosthetic "hand" that it used to pick up and bring food to its mouth. In 2005, a man was fitted with a thought-controlled bionic arm developed by a research institute in Chicago. Since then, at least fifty other people have acquired bionic arms. In 2012, a man with the first mind-controlled bionic leg climbed the 103-story Willis Tower in Chicago. It doesn't take a whole lot of faith or imagination to note that this could quickly move into Superman and Wonder Woman territory.

Translating thought directly into messages ("synthetic telepathy") was first achieved in the 1960s, when EEG was used to communicate in Morse code. The Massachusetts company Cyberkinetics now has BrainGate, a neural interface which allows the user to control a personal computer. Researchers at Duke University have created the first direct brain-to-brain interface between two rats.


## NOT JUST ELECTRONICS

Harvard Professor Takao Hensch used valproic acid to teach a group of adults perfect pitch, an ability never before known to have been acquired outside of childhood. A pill that restored plasticity in adult brains and allowed enhanced learning would be a tremendous breakthrough.


## TRY IT YOURSELF

Several companies are marketing affordable, non-invasive BCIs for neurogaming. These use EEG, EMG, or both to control aspects of games with brain waves, heart rate, and emotions. Various open source projects such as OpenEEG are devoted to neurotech, and a company called Backyard Brains sells inexpensive neuroscience tools.


## FARTHER OUT

Future BCIs might be amazingly capable. Researchers at UC Berkeley have proposed "neural dust": thousands of sensors just one hundred micrometers across that would be implanted in the brain and communicate via ultrasound with an implanted transceiver and an external device. It would be like having an MRI running all the time, and would enable continual measuring of brain activity, mental control of computers, and synthetic telepathy. Japanese researchers are working on a nanoscale electrochemical switch that mimics important functions of the human brain. Such an artificial synapse could be used to repair a human brain, or even in an artificial brain capable of hosting an uploaded mind.

Neurotechnology promises a better future, at least if we can avoid the secret government MKUltra-style mind controllers.

# Neural Conduction Sensors

**James Kent:** Neural conduction sensors monitor electrical activity through the skin, through the nerves, or through muscles. The most common form of neural sensor is the transdermal electrode array, a grid of pins stuck directly into the skin, nerves, and muscle tissue, much like you would stick a microchip into a circuit board. Kevin Warwick famously used a transdermal electrode array embedded in his forearm to control a bionic hand and share non-verbal neural pulses with his wife, who was also connected to a transdermal electrode array.

Even though transdermal sensor arrays require sticking pins into the skin, they allow a very fine level of control. Embedded electrode arrays are crude solutions for testing neural interfaces, and organic problems such as infection and scarring make them problematic for long-term use.

A non-invasive alternative to embedded electrodes is being explored by Ambient Technology in a device called the Audeo Sensor. The Audeo was developed to measure energy conduction around the laryngeal muscles for controlling speech production, and uses stainless steel electrodes to read differential voltage across the skin in the front of the throat. A user with the Audeo strapped to their neck can activate a phoneme-recognition dictionary just by thinking clearly in sub-vocalized speech . . . in other words, thinking about speaking. Using this technique, Michael Callahan of Ambient made the world's first voiceless phone call at the 2008 Texas Instruments Developer Conference in Dallas, using an Audeo sensor to "think" through a Bluetooth connection and have it translated to speech on his cell phone.

With the laryngeal muscles, there is a precise predictability to the rhythm of speech, making the task of aggregating sub-vocalized impulses at the muscle easier than capturing a similar brainwave library.

There is a subtle difference between thinking internal thoughts and producing sub-vocalized speech, but that difference is minor enough to overcome with some training. Noninvasive dermal contact plates, like those used in the Audeo, can be applied to any muscle group for exerting sub-motor control over external devices just by thinking about moving.

## Next Generation Interfaces

**James Kent:** Early experiments have demonstrated that neural tissue can be quickly adapted to communicate through embedded digital sensors. Based on an experiment with a monkey, mastering control of an embedded device takes days, not weeks or months. Training with embedded interfaces drives neuroplasticity and new signaling pathways to promote fast and robust connections with the device. Just as motor pathways grow to promote more precise control over machines, sensory pathways will grow to precisely capture, route, and analyze signals from sensory prosthetics. Neuroplasticity ensures that human networks can adapt to digital sensors through training, but currently digital sensors cannot adapt to human physiology.

Embedded sensors are not organic, so they can cause infection, scarring, and are susceptible to slow corrosion in an organic environment. Because of these limiting factors, even well-designed synthetic sensors will have a useful life of a few months to a year before scarring or slow degeneration means they must be removed or replaced.

Kevin Warwick, the man who embedded an electrode array into his forearm to control a robotic hand, has also demonstrated that rat neurons can be adapted in vitro (in a dish) to interface with a circuit board and control a small robot car. Taking this experiment from the motor to the sensory level, it is possible that a similar neuron-chip wired into a sensory feedback loop with a digital camera and video output electrode array will learn to use the camera and start navigating via visual cues.

## Fuckin' Wid Your Head

**James Kent:** While the primary purpose of neural interface research is putatively therapeutic, the functional potentials and ethical concerns of neural porting are problems looming in the future. Right now these are hypothetical concerns, but if a single-access embedded neurode procedure could be perfected and automated and performed at a local clinic in two hours for around a thousand dollars, and it was covered by insurance, the temptation for cosmetic and personal use of such a procedure becomes clear. Neural interfaces can be abused, obviously, and can be hacked into to enslave and torture minds, or drive people intentionally insane, or turn them into sleeper assassins or mindless consumers. Security is an inherent problem of any extensible exo-cortical system that must be addressed early in the engineering and testing stages, or anyone with an exo-cortical input would be ripe for exploitation. Sensory discrimination is an ongoing problem in any media environment, so individual channel selection, manual override, and the ability to shut down device input should be an integral part of any embedded system.

## Neural Interface for Increasing Intelligence

**James Kent:** How do we get people to become more intelligent within a single generation? There are a few popular answers to this question. The first is that humans take advantage of brain-computer-interfaces (BCI) to create more robust "offsite" memory and logic processing in a small microchip we keep implanted in our chest or shoulder. The technological foundation for making this work exists today, and is currently used to effectively treat Parkinson's disease via targeted computer stimulation of dopamine neurons. While the BCI option seems optimal at first pass, the fact that it requires surgery to embed electronics and pass dozens of thin electrodes into our brains at various areas presents ethical roadblocks to research. Perhaps if someone could finagle a sweet big-money grant to cure stupidity via microchip-aided neural synchronization, we would see some major progress in this area, but that's not likely in the US any time soon. Maybe China? Maybe India? Hello, developing world, I hear opportunity calling.

## Hearing to See

**Alex McKeown:** A Dutch physicist, Eindhoven-based Peter Meijer, has developed a device called vOICe that allows blind people to see, using the aural information entering their ears. The device converts visual information received by a camera into soundscapes, which then gives an aural representation of what the camera sees. The soundscapes are fed into the ears and—with training—the user can learn what visual information the soundscapes represent about the world. Exploiting the plasticity of the neural connections in the brain, the device tricks it into thinking that it is receiving information from the eyes. Consequently, the blind users see a visual representation of what is around them.

## Transcranial Direct Current Stimulation

**Ben Goertzel:** Transcranial Direct Current Stimulation (tDCS) is a relatively simple device that runs a very small direct current through the brain and can result in distinct, and sometimes significant, shifts in experience. The idea of running electrical current through the brain probably deters most folks and that is probably a good thing. Though the technology as it's intended to be used is extremely safe, used incorrectly it can obviously be dangerous.

# NUTRACEUTICALS

When the dreams of extreme longevity began to be popularized in the 1970s, lots of folks didn't want to wait for the pharmaceutical industry to come up with longevity drugs. They began researching and imbibing vitamins and nutrients. Durk Pearson and Sandy Shaw became the best known advocates for taking serious quantities of vitamins and nutrients to potentially slow down aging, increase intelligence, grow hair, get horny, and more, and formed a business that sold products that claimed to fulfill some of these hopes. Among those who became interested was the FDA, which didn't want these businesses making unproven claims about their products. Snake oil was mentioned. (Maybe snake oil works. Wouldn't that be funny?)

Since those times, these products have come to be defined as nutraceuticals. Debates rage endlessly about their efficacy. Ray Kurzweil takes a ton of them. Aubrey de Grey takes few, if any. Meanwhile, research and experimentation continue.

## RESVERATROL

I can remember doctors recommending a glass of wine every day to prevent heart attack way back in the 1970s—a piece of advice that is generally welcomed more readily than, say, a suggestion that one performs two hours of daily aerobic exercise.

Resveratrol is the stuff in red wine that is believed to make drinking to your health a sensible thing. Over the last decade or so, resveratrol has been shown to extend the lifespan of a type of worm, a type of fly, and a type of fish. Its antioxidant properties are believed by some medical experts to decrease the risk of cancer, heart disease, Alzheimer's, and diabetes. Tests appear to have confirmed that the effects of resveratrol are the result of the activation of a particular gene: Sirtuin 1. It is believed by some that caloric restriction provides a similar effect on the Sirtuin genes.

## CONFIRMATION?

According to a 2013 article in *Science* magazine, researchers at Harvard Medical School confirmed that resveratrol does provide anti-aging benefits.

David Sinclair, an Australian biologist and genetics professor, has become a leading voice for the impact of Sirtuin 1 as a key to slowing down diseases related to aging. He cofounded Sirtris Pharmaceuticals in 2004 to develop drugs based on this theory. The company was gobbled up (and then shut down) by the big pharma behemoth GlaxoSmithKline.

Resveratrol was also prominently featured in Barbara Walters' TV special "Live to 150. Can You Do It?" Since then, it's become one of the mainstays of the vitamin industry though many experts say the dosages sold by most companies are too low to have much of an impact.

## GENESCIENT

Genescient is a company that hopes to develop substances that can slow or retard aging by using advanced genomics. They have a particular focus on studying long-lived animals.

In 2006, they acquired the use of the genomics of the so-called Methuselah fly—a fruit fly that was bred for longevity and was made to live three times longer than average.

They claim that certain "designer supplements containing nutrients made using detailed genomic information"—a field they call nutrigenomics—will be able to slow down aging very soon.

In 2010, Genescient released STEMCELL100, which includes marsupsin and pterostilbene (resveratrol analogs). They claim this is the only supplement proven to double maximum lifespan of an animal model. No other product or therapy, including caloric restriction, even comes close.

In late 2013, Genescient came out with a genetically engineered memory supplement called Memex 100™. Memex 100 was developed by screening various substances on flies carrying human genes for Alzheimer's. The final winning formulation of eight substances was able to essentially block the effects of human Alzheimer's genes on fly coordination. Memex 100 is currently being sold as a dietary supplement to support memory, but it is also undergoing clinical trials with Alzheimer's patients that have mild to severe Alzheimer's disease.

# OPEN SOURCE

"Open source" is a term that's widely used within digital culture. The term was coined by Christine Peterson (see Foresight Institute) in 1998, but the concept originates from early computer hacker culture. The convention was to make source code available so that other hackers could add to it, alter it, and use it any way they chose.

Today, after the evolution of huge commercial markets in computer programs, open source refers to software designs that are freely distributed and are open for alterations. Open source is the opposite of the proprietary software produced by most computer companies, including Apple and Microsoft.

## OPEN SOURCE EVERYTHING, INCLUDING TRANSHUMANISM

Open source has, in recent years, seen wider usage as a description of all sorts of projects in a variety of fields. As we've already seen in our discussion of citizen scientists, there's a great passion for open source activities within the transhumanist world. This is, perhaps, a saving grace for advocates of the sorts of technological enhancements that so much dystopian science fiction imagines being completely controlled by massive corporations. Open source levels the playing field and opens up space for DIY projects to gain both expertise and technological influence.

## Proprietary Knowledge Dies, Open Source Knowledge Continues

**Dr. Joel Pitt (interviewed by Joseph Jackson):** More than one commercial project with AGI-like goals has run into funding problems. If the company then dissolves, there will often be restrictions on how the code can be used or it may even be shut away in a vault and never be seen again. Making a project open source means that funding may come and go, but the project will continue to make incremental progress.

## Intellectual Property Laws Threaten Progress

**David R. Koepsell (Interviewed by Joseph Jackson):** I've been researching and writing about confusion and paradox in intellectual property law for the past fifteen years . . . I am interested in how patenting threatens the early stages of nanotechnology and synthetic biology. I have also recently started to argue that there is a fundamental ethical problem with intellectual monopolies. I believe that the future of both science and innovation hinges upon rejecting IP as it has been traditionally construed (for both ethical and pragmatic reasons), and that open source, private law models will govern more efficient exchanges in the future.

## AGI Transparency Is Against Elites

**Dr. Joel Pitt (interviewed by Joseph Jackson):** I'm completely against the idea of a group of elites developing AGI behind closed doors. Why should I trust self-appointed guardians of humanity? This technique is often used by the less pleasant rulers of modern-day societies: "Trust us—everything will be okay! Your fate is in our hands. We know better."

The open source development process allows developers to catch one another's coding mistakes. When a project reaches fruition, it typically has many contributors and many eyes on the code can catch what smaller teams may not. It also allows other friendly AI theorists to inspect the mechanism behind an AGI system and make specific comments about the ways in which unfriendliness could occur.

When everyone's AGI system is created behind closed doors, these sorts of specific comments cannot be made.

**Misha Angrist (interviewed by Joseph Jackson):** I see open science as a welcome antidote to a century of top-down management of human biology characterized by condescension, secrecy, and unintended consequences. The resistance of the medical establishment to personal genomics—as manifested by calls for randomized clinical trials, stringent regulations, and steadfast, if naïve, resistance to patient and consumer activism—leads me to believe that I'm right. We are in the midst of a revolution, yes, but the outcome will not be dictated by the medical-industrial complex, just as the future of journalism will not be dictated by the purveyors of the daily newspaper.

# OPTOGENETICS

*(See also Cognitive Science, Cognitive Enhancement, and Neurotechnology)*

## BETTER BRAIN CONTROL

Humans are always trying to control brains, either their own or other people's. Most methods are indirect: education, propaganda, entertainment, meditation, and drugs. Direct control (neuromodulation) is more precise but more difficult, involving electricity, magnetism, or chemicals.

One type of neuromodulation is deep brain stimulation (DBS), which implants electrodes in the brain (a "brain pacemaker") for treating disorders such as Parkinson's disease or chronic pain. But as advanced as DBS might be, it's still rather crude to activate or suppress brain activity by using wires to zap millions of cells.

## LIGHTING UP YOUR NEURONS

Optogenetics is still brain surgery, but far more precise. Start with algae that have proteins that respond to certain wavelengths of light by allowing electrochemical ions to pass in or out of the cell. Take the gene for that protein and insert it into the DNA of a specific neuron type in the targeted area of the brain. The inserted genes make those neurons responsive to light. Add a fiber optic cable thinner than a human hair and a laser, and you have a very precise on/off switch, even for areas too deep or fragile for other types of brain surgery.

## TIMING IS EVERYTHING

Optogenetics is powerful not only because cells can be targeted with improved spacial precision, but also vastly better temporal precision. Neurons fire on the millisecond timescale, and a slight timing shift can change or even reverse the effect of the signal. By operating on the same timescale, optogenetics can add or delete precise electrical patterns in specific cells, in realtime.

Imagine operating a soundboard during a live musical performance, adjusting the mix by boosting or suppressing the signals from different microphones and at different frequency ranges. With your "optogenetic soundboard," you would have control of individual "notes."

Optogenetics is already revolutionizing neuroscience, helping us to understand OCD, vision, memory, hunger, addiction, aggression, and depression, and to treat neurological disorders. It also looks like a key technology for enabling future neurological enhancement.

# PERFORMANCE ENHANCEMENT

If enhancement is the main trope of transhumanism, then performance enhancement is the place where we get to witness if this enhancement thing is working. After all, it'll take about one hundred and twenty years to find out if those longevity drugs are really going to let a thirty-year-old live to one hundred and fifty, but we quickly find out if some drug helps an average student boost his or her GPA, or if some other substance makes Joe Slacker into a long-distance runner. Performance enhancement appeals to that competitive, American pursuit of excellence, but the competitive aspect of it may become meaningless if everybody is enhanced. Is excellence for its own sake enough?

Controversy over performance enhancement has focused largely on sports, with various sports heroes like Barry Bonds and Lance Armstrong being denounced and censured by officials and fans. On the other hand, no one has yet suggested that someone's law degree should be revoked because he or she used Adderall to pass the exams. And no one has suggested that *Sgt. Pepper's Lonely Hearts Club Band*, often cited by critics as the greatest rock album ever, should be demoted because The Beatles were using a lot of LSD when they made it.

## Game Enhancement vs. Life Enhancement

**Quinn Norton:** Elite (sports) players get it all: performance-enhancing drugs, surgeries, gadgetry, specialized equipment, even mathematical analysis to help them perform their desired tasks. They are monitored and modeled, tested and retested, sorted and classified. The modern elite player is an isolated cyborgian construct with barely room for a life and identity away from their sport.

Current attitudes toward enhancements vary wildly. Some enhancements are considered the price you pay to get in the game; others, the worst type of cheating.

This may seem hypocritical, but it isn't. After all, the rules of sports are arbitrary. Why shouldn't you use your hands in soccer? Because then it's not soccer. What makes a hypobaric chamber okay, but an injection a firing offense? Because we say so. After we invented agriculture, the bow, or perhaps mountaintop mining equipment, human athletics became a cultural pastime rather than a vital function.

This is why applying the debate about sports enhancements to the rest of the world can be dangerous. When we're deciding if we should give modafinil to pilots or Ritalin to grad students, we're making life and death choices about what our future will look like.

# PETER THIEL

Peter Thiel is a controversial billionaire venture capitalist and hedge fund manager who has lent his support to a fair number of transhumanist and singularitarian ventures. Thiel made much of his change as the co-founder of the money transfer site PayPal and as an early investor in Facebook.

Thiel is an avowed believer in The Singularity and has given substantial financial support to The Singularity Institute for Artificial Intelligence (now the Machine Intelligence Research Institute) and to Aubrey de Grey's Methuselah Foundation and SENS Research Foundation.

Thiel is radically libertarian. An article he wrote for the Cato Institute's journal *Cato Unbound* in which he declared, "I no longer believe democracy and freedom are compatible," and "the extension of the franchise to women" has "rendered the notion of capitalist democracy an oxymoron," led to a bit of an informal split in transhumanist circles, in which some on the transhumanist left are less inclined to collaborate with other transhumanist groups.

# POST-DARWINIAN

*(See also Abolitionism, Designer Babies)*

The notion of a post-Darwinian epoch, as it's employed by transhumanist sorts, refers to a future time when the desires and behaviors of humans (or posthumans) will no longer be driven by the dynamics of biological evolution. Biological Darwinian evolution was famously described by the 19th century English poet Alfred Lord Tennyson as "red in tooth and claw" for the way in which life feeds off of other life. At its extreme, post-Darwinians like David Pearce would have all lifeforms relieved from the struggles of Darwinian life.

## No More Unpleasantness?

**David Pearce:** Over time, I think . . . unpleasant states of consciousness—unpleasant states that were genetically adaptive in our ancestral environment—will be weeded out of the gene pool. A very different kind of selection pressure is at work when evolution is no longer "blind" and "random," i.e. when rational agents design the genetic makeup of their future offspring in anticipation of its likely effects. In that sense, we're heading for a post-Darwinian transition.

# POSTHUMANITY

According to some thinkers, transhumanist technologies will give birth to the posthuman, a creature distinct enough from us to merit that name. One supposes that this would depend on the definition. Is an embodied being with a possible lifespan of thousands of years posthuman? What if he or she has a hyperintelligent AI implant supplementing and (in all likelihood) overtaking his or her biological brain? And what if said person is uploaded into digital worlds?

Another version of posthumanism would have it that our AIs will simply *be* the posthumans—leaving us in the evolutionary dust, either as a doomed species or enjoying us as we do our pets.

**Kyle Munkittrick:** The three flavors are critical posthumans, transcendent posthumans, and transhumans.

*Critical Posthumans:* This is what we all are. Critical posthumanism is simply the idea that our concept of "human" as a natural, non-technological thing was wrong from the beginning. Humans are most human when using technology, modifying ourselves and our surroundings.

*Transcendent Posthumans:* This is what we (most of us) wish we were. Flawless, immortal, godlike. Abilities so above and beyond humans that they are almost unimaginable.

*Transhumans:* This is what we are becoming. Nanotech, organ transplants, genetic engineering, prosthetics, cognitive and mood-enhancing drugs, cloning, morphological freedom, and anti-aging medicine are a small sampling of the tech helping us overcome our biological limits.

# POST-SCARCITY

Post-scarcity is a vision in which human economic relations are no longer guided by a scarcity of things that are necessary—or perhaps even desirable—for human survival and fulfillment. Some visions, like those of the eco-anarchist Murray Bookchin, while based in the ability of technology to provide basic livelihood, advocate—and rely on—shifts in lifestyle in which consumer values and desires are abandoned in favor of more bucolic and uncommercialized social relations, generally under a sort of libertarian form of voluntary cooperation and sharing.

## EVERYTHING ALL THE TIME

Developments such as NBIC seem to some to promise a more indulgent form of post-scarcity in which our desires for stuff, pleasures, and extreme experiences can be sated, if that's the way we choose to live, even with little or no health or environmental downside.

## AN UNLIMITED SUPPLY

The notion of post-scarcity has historical ties mainly to leftist politics (Karl Marx believed socialism would bring it to fruition). In more recent years, thinkers of varied stripes have embraced the idea that abundance could result from technological progress, including many who believe that the free market is the best way to get there.

Some claim that a sufficiently post-scarcity situation (for example, if everyone has 3D printers that can cheaply make anything they want) will radically reduce the importance of supply and demand in market capitalism.

## FUTURE SCARCITIES OR FUTURE ABUNDANCE?

Peter Diamandis, Chairman and CEO of the XPRIZE Foundation and co-founder and Chairman of Singularity University, wrote the popular recent book *Abundance: The Future Is Better Than You Think*. To the credit of Diamandis and his co-author Steven Kotler, the book doesn't merely offer vague promises of a cornucopia once the tech gets good enough. It actually looks at numerous areas (potable water and energy, for example) where we may, in fact, be facing a crisis of *scarcity*. The authors then discuss current nascent tech developments, as well as specific projects, that might resolve and even surpass each threatened scarcity, bringing about the longed-for abundance. Diamandis and Kotler do such a good, honest job of covering the areas in which we face big scarcity challenges that some thoughtful readers may actually find their premise less convincing after reading the book.

## HOW DO WE KNOW IF WE'RE IN POST-SCARCITY?

In some sense, those of us living in technologically advanced countries are already in a post-scarcity situation. Few of us, for example, seem to be needed to produce or distribute essential goods or provide essential services, even if we expand our definition of essential to include things like transportation and communications technologies. This situation is one of the factors that has led some to call for a guaranteed income. However, the situation is complicated by a number of factors, among them our reliance on products from poorer parts of the world, cheap immigrant labor, and a number of vital areas in which scarcities continue to exist.

## A FIRST STEP TOWARD THE ABOLITION OF SUFFERING?

As we learned early in this tome, some particularly idealistic transhumanists want to bring an end to suffering in humans—and some even want to end it for all sentient beings. For these transhumanists, ending scarcity is at, or near, the top of their list of priorities. Aside from the suffering caused by poverty, resource scarcities can lead to conflicts and wars, and warfare spreads suffering.

> **Valkyrie Ice:** You may have heard of the term "scarcity economy," and its opposite, the "post-scarcity economy." In brief, in our current economy, value is determined by how "scarce" any given product is. But if your printer can print just about anything, from food to electronics to household furniture—then what can be defined as "scarce"?

## A Utopian Post-Scarcity Scenario

**Jason Stoddard:** Let's be clear on this: we're not going to wake up in a magical world where iPods and McMansions grow on trees overnight. Before that can happen, every part of today's value chain has to be overturned. Everything. Production of raw materials, transport and refining, design and engineering, manufacturing, distribution . . . even our own sense of worth. So, if today's financial crisis is the first step, where do we go from here?

We're already starting to see some examples of near post-scarcity. Consider computers and communications. If you're willing to use a computer that's a couple of years old, you can probably find a hand-me-down for free, and then happily talk to your friends around the world on Skype using free public wi-fi.

Or consider that in the last Depression, the main worry was simply getting enough food. Today, the marketplace is more worried about maintaining the marketing budgets of 170 different kinds of toothpaste than about ensuring that everyone has toothpaste. There's a lot of padding in the system. Couple a financial crisis with this overweight, inefficient system, and you have the stage set for the first transition to post-scarcity: a comprehensive rethink of our concept of value.

*Transition Phase 1—Value Proliferation:* Today, rappers sing about driving Bentleys, living in hotel-sized mansions, and drinking thousand-dollar bottles of cognac. Soon, they may be saying, "And that don't mean shit unless you got viz and virt and rep!" We've already seen the beginning of this: divorce cases in which World of Warcraft's internal currency is named as an asset;

the growing importance of reputation systems such as eBay feedback; the proliferation of corporate "points" or "bux" systems that can be exchanged for real goods; the monetization of attention via friend-spamming on social networks and advertising on popular blogs. Our concept of value is expanding; it will expand even more in this phase.

Think about it. If real currency, virtual currency, corporate points, visibility, and reputation all have value, exchanges will soon crop up. Think of a FOREX (a market in which foreign currencies are exchanged) for all things we consider of value. As point examples of near post-scarcity grow and these value systems become interlocking, we'll move beyond a single monetary value system. You'll be able to live well under any number of value systems: reputation, visibility, network, rewards points, or even "old-fashioned" currency. *(Ed.: Bitcoin emerged after Stoddard wrote this, but it certainly is playing a role in expanding and confusing value proliferation.)*

*Transition Phase 2—Unseen Golden Age:* The next phase of the transition to post-scarcity is the scariest, but only if you look at it from today's POV. What's hard to accept? Well, multiple interlocking value systems require comprehensive metrics and tracking. Read: surveillance. We could easily find ourselves in a propagational economy, where a person's entire value is based on their Attention Index (their visibility to other people) and Monetization Effectiveness (how well they sell).

"Yuck," you say.

But what if advances in manufacturing efficiencies make it possible to live well, simply by interacting with friends and going about your life? What if below-replacement-level birthrates and advances in biotechnology mean you could check out of the system by claiming a piece of unused desert and planting a house? This surveillance economy might be a very easy place to live.

The end of this phase would come rapidly if Drexler-level nanomachines (molecular manufacturing) made the production of material stuff essentially free, and took the future worth of the entire value chain to zero. If it costs nothing to make the machines to find and refine the raw materials, or to grow the transportation network, two of the "insurmountable" obstacles to post-scarcity disappear. Even without this near-magical technology, bioassembly and other methods will slowly erode the value of raw materials refining, transportation, and manufacturing. In either case, this is an even bigger economic rethink than the one we're going through today.

*Transition Phase 3—Magical Ideas:* True nanotech is limited only by the energy we put into it. In this time, unthinkable mega-engineering projects become feasible, for example, growing a global network for finding, refining, and transporting raw materials; producing hundreds of space elevators for easy access to extraterrestrial resources; assembling magical factories along every coastline. In this phase, we are truly free to dream and big ideas are the currency. The dreamers and designers who can imagine the best ways to change the world will become the "economic" giants of their time. The big issue will be how to coordinate these visions, and to eliminate or minimize disruptive ones.

This phase ends when the systems for effortless production of all our dreams are in place. Artificial intelligences or powerful semantic processing make this unlimited capability accessible to anyone. We are now free to imagine what we want—and have it delivered on demand.

True post-scarcity means speaking your wishes to the air and having them delivered. The seamless nanotech and biotech skeins distributed through the Earth and the solar system make every wish possible. The only remaining question: where do your rights end and someone else's begin?

Now, sit back and think. Even without life extension, I might see every phase of the transition to true post-scarcity in my lifetime. And remember that thought the next time you check your brokerage account. It is the end of the world as you know it. And that is perfectly fine.

# PROACTIONARY PRINCIPLE

## FIRST, THE PRECAUTIONARY PRINCIPLE

Max More developed the Proactionary Principle in opposition to the Precautionary Principle—a fairly ubiquitous although rarely enforced idea that has been endorsed by the UN, the EU, and others. The Precautionary Principle states that a technological development or public policy should not be employed unless there is proof that the development will not cause harm to the public or the environment. Of course, it's impossible to prove this with anything close to certainty, so a literalist interpretation of the principle would mean ceasing most productive activities.

## PROACTIONARY

More's Proactionary Principle suggests that we should consider both the risks and benefits of developments and activities, and we should also consider the risks of *not* doing certain things. In Version 1.2 of the Proactionary Principle, More states, "If the precautionary principle had been widely applied in the past, technological and cultural progress would have ground to a halt. Human suffering would have persisted without relief, and life would have remained poor, nasty, brutish, and short: No chlorination and no pathogen-free water; no electricity generation or transmission; no X-rays; no travel beyond the range of walking." Someone in a future transhumanist society might say: "If the precautionary principle had been widely applied in the past, humans would still suffer from poverty, illness, stupidity, and death." Or they might say, "Damn. I wish we'd never let the Kardashians get hold of those boob-growing nanites" while drowning in a sea of human breasts.

Seriously though, unless the precautious ones find a way to modify their principle so that it's not absurd, Mr. More is the reasonable one here.

# PROSTHETICS

*(See also Exoskeletons, Neurotechnology)*

As a technology that is already showing signs of evolving from helping people cope with handicaps to actually improving the performance of various limbs and organs, prosthetics are, perhaps, the populist edge of transhumanism.

## THE KILLER CYBORG

The most popular reference point for prosthetic enhancement was South African sprint runner Oscar Pistorius, a double leg amputee whose speed and agility using prosthetic legs allowed him to win various championship races, in some cases competing against people running on their original limbs.

There was controversy over whether his prosthetic legs gave him an unfair advantage, and his request to compete in the 2008 Summer Olympics was rejected. He was allowed to compete in a number of races in the 2012 Summer Olympics. He didn't win, but he did come in second place in a 400-meter race.

In 2013, Pistorius was arrested and charged with shooting and killing his girlfriend, tarnishing his brand and complicating his value as an example for enhancement advocates.

## WAITING FOR REPLACEMENT

Today, we appear to be on the cusp of a revolution in remaking the human body. We have artificial replacements for nearly any limb, joint, or internal organ.

As scientists work, for example, toward making an artificial eye that can see as well as the one it replaces, it becomes clear that seeing even better yet may simply be a matter of an exponential growth in microchip technology. Of course, there's the question of the brain's ability to adapt to a different sort of visual data, but as we've seen, we're already looking at neural implants and other methods for enhancing the brain.

Some envision not-too-distant future humans having had most—or all—of their body parts replaced by improved artificial ones, including that center through which we apprehend the world (no, not the crotch!), the brain. The question becomes whether the totally cyborgized individual is still the same individual.

# PSYCHEDELIC
# TRANSHUMANISM

Whence comes inspiration? Neurology has enough problems figuring out the manifold automatic processes operating beneath the realm of conscious thought; or our multiplex deductive reasoning; or the source of our self-awareness. What are we, then, to make of a heightened—sometimes ecstatic, sometimes fearful—sense of melting into or merging with all life, or the sudden and thrilling emergence of insight and perspective unbidden, not resulting from a chain of logical reasoning?

While some prefer to believe that such inspiration and perspective shifts come from a divine source, we are increasingly learning about how prosaic alterations in neurochemistry can lead to such poetic peak experiences. (We leave open the possibility that the potential for these neurochemical changes was granted by some sort of divinity.)

While some aspects of these heightened experiences and radical perspective shifts may happen naturally or even, occasionally, from the ingestion of certain mundane foods—or alcohol, opiates, or stimulants—psychedelic chemicals and plants virtually guarantee a radical perspective shift. And while the experience often produces nothing more than confusion or an overwhelmed blissful contentment with listening to Brian Eno's *Music for Airports,* those who use these mind-altering agents with a modicum of expertise (or just plain luck) will likely experience a heightened sense of pattern recognition, novel insights, or a subtle sense of having been refreshed, with the resulting "flow" lasting for days, weeks, or months after.

For these reasons, many—probably even most—transhumanists think of psychedelic drugs as mind enhancers.

## PSYCHEDELICS AND TRANSHUMANISM

The relationship between psychedelic culture and transhumanist culture is long and complicated. Perhaps the first pop-culture figure to enunciate a set of transhumanist goals was the "LSD guru" Timothy Leary, who suggested Space Migration Intelligence Increase and Life Extension as goals for a more expansive human condition in 1974. (See entry for Timothy Leary.)

Max More's "Principles of Extropy" embraced the use of neuro-chemical enhancement, and in further exploration of that topic, Mr. More included psychedelic drugs among the enhancers.

On the other hand, many transhumanists concern themselves with increasing human rationality, and find psychedelics and the claims made around them suspiciously subjective. And many of those actually working in the various fields that are creating the transhuman condition simply find them irrelevant. In a reversal of the cultural dynamics of the "psychedelic '60s," younger transhumanists are more likely to be contemptuous of psychedelics and their efficacy.

## WHOSE PERFORMANCE WAS ENHANCED?

Sir Francis Crick was helped by low doses of LSD in his role in discovering DNA. Steve Jobs and many other digital pioneers were inspired and assisted by psychedelic drugs (as covered by the book *What the Dormouse Said: How the Sixties Counterculture Shaped the Personal Computer Industry* by John Markoff, and *Steve Jobs* by Walter Isaacson). Transhumanist leader and singularitarian roboticist Ben Goertzel is open about his use of psychedelics for inspiration, as is Mark Pesce, who coded VRPL (Virtual Reality Programming Language). Early life extension hypesters Durk Pearson and Sandy Shaw were also vocal psychedelic enthusiasts. And then there are the dozens who would rather not say, due both to the illegality of psychedelic substances and the skepticism in officious scientific circles regarding insights gained from these experiences.

**David Pearce (interviewed by Michael Garfield):** I think it's fair to say the transhumanist community is mostly interested in intelligence-amplification—superintelligence rather than supersentience. I share an interest in cognitive enhancement, but in my opinion there is an important sense in which a congenitally blind person with an IQ of 220, or 920, is just as ignorant as a congenitally blind person with an IQ of 120. I worry more about our ignorance in the latter sense than I do about our limited reasoning powers. Psychedelic drugs can briefly give us a tiny insight into how "blind" we normally are, but we soon lapse into ignorance again. Such is the state-dependence of memory. If I'd never tried psychedelics, then I fear I would be scornful of their significance because of the incoherence of most users' descriptions of their effects. But using the blindness analogy again, someone congenitally blind who is surgically guaranteed the gift of sight can take years before they can make sense of the visual world . . . at first they are overwhelmed and confused by visual stimuli.

**Michael Garfield:** Some transhumanist visionaries have been catalyzed by the psychedelic experience. They understand the message of psychedelics and the message of technology to converge on the horizon of a deeper reading of reality that recognizes mind and matter as dimensions of the same truth—a truth for which language has ill prepared us.

Among the ranks of these "psychedelic transhumanists" are legendary rebels like Timothy Leary, wise fools like Terence McKenna, cultural commentators like Erik Davis and Mark Pesce, and avant psychopharmacologists like David Pearce.

Their common vision shares much with the rest of the transhumanist community, including an embrace of technology and science as both potent and inevitable; an evolutionary model of the universe and humanity; a sense of the human organism as something that can be tinkered with and expanded; a recognition of drugs as a technology that can dramatically reinvent identity; and a participation in a playful challenging of fixed boundaries. In many ways, they demonstrate the seed of transhumanism in this moment by exemplifying self-revision and the reevaluation of assumptions as an open-ended and ongoing process. And along the way, they tatter the mechanistic control fantasies we have held onto in spite of our most sophisticated inquiries.

These visionaries . . . tend to critique philosophies that consider mind a mere epiphenomenon, or that fail to recognize the role of the speculator in speculation.

They see technology as ideas, and ideas as technology. They question our fanatical efforts at control . . . and remind us of the stubborn persistence of the unconscious, the body and the other. They remind us to see the evolution of humanity and beyond as much in terms of qualia as quanta, and paint the future as more sensitive to psychological, spiritual, ethical, and biological concerns than those on the hardboiled tech edge.

**Michael Garfield:** If one thing makes itself apparent from the psychedelic experience, it's that the more you know, the more you don't know, and admitting this is a form of death. The acceleration of intelligence and extension of the individual lifespan means that life itself will increasingly come to resemble a constant re-imagining of self—not the indefinite perpetuation that many of us desire, but an ongoing process of death and rebirth. And by its very nature, death is across the event horizon, an impenetrable unknown.

# QUADCOPTERS

*(See also Sousveillance)*

## THAT'S NOT A BIRD

They may seem to be merely the latest versions of the radio-controlled flying toys that have been around for decades, but quadcopters are a lot more than that.

Their physical form is deceptively simple: four propellers at the ends of an X-shaped fuselage. Each diagonal pair of blades rotates in the opposite direction, eliminating torque-induced yaw and thus the need for a tail rotor. Most quadcopters are radio controlled, but some are controlled from a PC or even a smartphone. Flight control is achieved by altering the speed or pitch of the rotors, which eliminates the need for changing individual blade pitch angles with the complex joints used in normal helicopters.

Add a tiny video camera (or more than one for stereo and panorama views), and perhaps some microphones and other sensors, and you have a highly maneuverable surveillance platform that one person can afford and operate. You've got aerial views on the cheap, and a device small and controllable enough to fly in an open window and land on top of a hanging fluorescent fixture.

## QUIETER, SMALLER OR LARGER, ALONE OR IN SWARMS

Quadcopters are rapidly evolving. Advances in batteries and fuel cells will enable them to fly longer. Noise-cancelling technology will make them nearly silent. Smaller ones will become ever stealthier, and larger ones will carry more payload. On-board software gives them the ability to fly autonomously, without an operator. With multicraft communication, swarms of them can fly in formation.

## SURVEILLANCE: IT'S NOT JUST FOR THE AUTHORITIES ANYMORE

Like most technology, quadcopter applications range from the trivial to the clearly beneficial and the sinister. They make a fun platform for games. They're a potentially valuable tool for crime prevention, search and rescue, scientific research, environmental sensing, remote telepresence, and augmented reality (AR).

Obviously the police and spy agencies see a lot of potential, but don't panic. The potential cuts both ways. The game changes when a spy gadget is no longer the sort of rare and costly item that Q would have handcrafted and given to James Bond, but rather something the average citizen can afford. When surveillance becomes democratized, it becomes sousveillance, the recording of activity by participants (see that entry).

## DEATH FROM ABOVE: PERSONAL EDITION

Now for the sinister. Soldiers already use small drones to safely take a look over the next hill, and, of course, the military would love to figure out how to weaponize them. The bigger-booms-are-better philosophy peaked with nuclear weapons, and in recent years the Pentagon has tried to adopt a less-is-more approach: more precise weapons that can get the job done cheaper, with less risk to the troops, and with less collateral damage. It's also much better for PR.

So they'd rather not kill a terrorist by sending a bomber to level a neighborhood. Today, a large drone (remotely controlled from possibly thousands of miles away) launches a missile and blows him up in his house or car. But even that seems crude and messy compared to having something that would fit in a briefcase, fly in through a window, confirm the target via a realtime video link or on-board facial recognition, and shoot him. Tom Clancy would have approved.

## SO WHAT'S THE TRANSHUMANIST ANGLE?

Quadcopters are another emerging technology that expands the reach of human senses and empowers individual human beings. Increased surveillance and sousveillance will lead to losses of secrecy and privacy, but gains in knowledge and accountability. And, hopefully, those autonomous assassination bots won't make mistakes or fall into the wrong hands.

> **Valkyrie Ice:** Small drones would enable nondisruptive observation platforms for nearly every environmental and anthropological science. Imagine a solar powered drone programmed to follow a wolf pack, or to investigate the upper levels of a rainforest. Imagine tiny drones able to follow the daily life of a bee or a hummingbird.

# QUANTIFIED SELF

Some of us would rather wait for technology to give us health and longevity than to engage, literally, in bean counting. However, for those who have a special love for monitoring their personal data and who want to maximize their wellness and performance, there's the new culture of the quantified self (originally called lifelogging).

Participants track what they take into their bodies, their physical activities, their moods, and the quality of their performances in daily activities. They often employ wearable technologies to measure things like heart rate, blood pressure, brain activity (using EEG), perhaps even fat-to-muscle ratio.

Gordon Bell, a widely noted researcher at Microsoft, is sometimes credited with starting the movement in 2004. With his MyLifeBits project, he began to digitize and store everything—from every webpage he looked at, to everything he wrote or read, his emails, his phone messages, ad infinitum. His colleague, Jim Gemmell, created software to automate many aspects of his lifelogging. Their 2009 book *Total Recall: How the E-Memory Revolution Will Change Everything* tells the story.

Scanadu is a technology inspired by *Star Trek* tricorders: a handheld device that is placed on your forehead and measures, saves, and time-stamps data including heart rate, skin temperature, core body temperature, $SpO_2$ (oxygen saturation), respiratory rate, blood pressure, ECG, and more. Projects like this promise to make the quantified self intuitive and effortless. Self trackers may be built into cell phones or watches by the time you read this.

**Alexandra Carmichael:** CureTogether is a patient data-sharing site I co-founded with Daniel Reda where people come to self-report symptoms, treatments, and triggers for over three hundred conditions.

People are tracking their depression, cholesterol, migraines, and countless other measures. Using migraine as an example, patients visiting CureTogether can see community statistics and learn that the top reported symptoms are "nagging pain in one side of the head" and "nausea"; the top reported treatments are "sleep" and "ibuprofen"; the top reported triggers are "stress" and "not enough sleep" and the top related conditions are anxiety and depression.

Instead of narrative websites that provide emotional support in the form of shared disease stories, the quantitative data at CureTogether enables decision support and hypothesis generation. People are getting ideas for new treatments that they ask their doctors about. They are seeing how common or rare their symptoms are and learning what triggers might be affecting them. While each individual's data is completely private, the aggregate data is open for researchers around the world to analyze and use to make discoveries for the greater good. Some interesting correlations are already starting to emerge, like a potential link between migraine and fibromyalgia.

## Quantified Self Exemplar

**Alexandra Carmichael:** It probably won't surprise my readers that I track myself. But it might surprise you that I track forty different things every day. On a typical day, my pain level is two, my weight is one hundred twenty-six pounds, I do one hour of walking, my happiness is nine, and I sleep six hours.

With a background in molecular genetics and bioinformatics, as well as a history of chronic pain, I started tracking to help myself. But I soon wanted to apply what I had learned to help others.

# QUANTUM COMPUTING

## PUTTING A STRANGE WORLD TO WORK

Standard computers use strings of binary digits (a one or zero), or bits, as their basic units of information. Quantum computers use quantum bits, or qubits, which can be one, or zero, or simultaneously one *and* zero. ("If you think you understand quantum mechanics, you don't understand quantum mechanics," said Richard Feynman, and he wasn't kidding.) In the quantum world, things like photons, electrons, atoms, and even molecules can exhibit superposition: being in multiple states at the same time. And then there's entanglement: a way quantum particles can remain linked, acting in unison, even when they are separated. Einstein called that "spooky action at a distance" (and he wasn't kidding, either). Quantum computers use superposition and entanglement to calculate, transmit, and store data. While not better than standard computers for everything, for some problems they are far faster.

## THIS IS NOT EASY

Precise control is difficult, because real-world qubits are very delicate. Premature decoherence (loss of their quantum state) stops the calculation and the data is lost. The lifespan of qubits is typically measured in minutes, and the more you use, the more difficult it becomes. Even the $10 million quantum computer from D-Wave Systems uses only 128 qubits. Still, customers like NASA, Google, and Lockheed Martin think it's worth the cost.

However, some scientists are skeptical about whether D-Wave's computers really work as claimed. They note that peer-reviewed papers show scientists struggling to make far fewer qubits work together, so

D-Wave's claims of commercial machines using 128 and even 512 qubits raise doubts.

## WHY IT'S WORTH IT

The fastest standard computer basically does only one calculation at a time. (Even with multiprocessing, each CPU or core can only do one thing at a time.) In contrast, a quantum computer can entertain the whole domain of a particular calculation simultaneously, and so is inherently faster at searching through a space of potential solutions to find the best one. This makes them well suited for many extremely complex computational problems.

If you are doing drug discovery and need to test trillions of combinations of amino acids to find one special protein, you'll want a quantum computer. At least, you will if you want the answer before the sun runs out of hydrogen. Other uses include validation of large software systems and efficiently simulating many-particle quantum systems. (That last one is crucial if you want to run your own universe, à la simulation theory.)

## CRACKING CODES . . .

Quantum computers are perfect for factoring the large numbers used in cryptography. Currently, such tasks can take hundreds of computers working for years, as it did in 2009 when researchers succeeded in prime-factoring a 768-bit encryption key. Breaking one of today's commonly used 1024-bit keys would have taken them hundreds of times longer, but all forms of public key encryption could be broken by a large enough quantum computer. So, as you might expect, the NSA is working in this field, though officials aren't talking.

## . . . AND MAKING BETTER ONES

Secrecy may seem doomed, but with something called quantum key distribution (QKD), Heisenberg comes to the rescue. Just create pairs of entangled photons, and send one of each pair to your distant partner. Because the photons are entangled, their polarizations will be perfectly correlated. With enough photons you can build up a key long enough to be secure. If someone intercepts your communications, Heisenberg's uncertainty principle kicks in: the snoop's observations would change the measurements, the keys would no longer match, and you'd know someone was listening in.

**Richard Loosemore and Ben Goertzel:** One objection to the idea of an AGI intelligence explosion is the idea that human-level intelligence may require quantum computing.

The working assumption of the vast majority of the contemporary AGI field is that human-level intelligence can eventually be implemented on digital computers, but the laws of physics as currently understood imply that to simulate certain physical systems without dramatic slowdown requires special physical systems called "quantum computers" rather than ordinary digital computers. But there is currently no evidence that the human brain is a system of this nature.

# RAPTURE OF THE NERDS

This singular and oft-used putdown of the singularitarians was first used publicly by Ken MacLeod in his 1998 novel *The Cassini Division* (although it may have been in nerd usage prior to that).

The Singularity does provide for many of the desires that were earlier answered by religion. Among these are eternity, transcendence of the flesh, heavenly transcendent worlds (improved Earth, space colonies, or digital space), and a payoff for keeping your nose to the productivity grindstone and staying out of trouble.

Most singularitarians, though, are atheists and bristle at the snarky phrase. In 2012, two uber-nerd favorite writers, Charlie Stross and Cory Doctorow, teamed up to write a science fiction satire titled *Rapture of the Nerds*. It's full of transhumanoid and singularitarian high concept (much of which you've read here), but—according to my alter ego Malcolm McLuhan—the book is merely amusing, not devastatingly hilarious. And I wouldn't argue with me.

## Oh Yes . . . We Forgot to Mention Resurrection

**Giulio Prisco:** "The Rapture of the Nerds" represents a moderately recurrent meme in certain circles—to denigrate transhumanism by comparing it to extreme religious notions. But not all transhumanists consider such comparisons wholly off-base. While transhumanism differs from traditional religions in being based around reason more centrally than faith, it does have some commonality in terms of presenting a broad vision of the universe, with implications on the intellectual level but also for everyday life. And it does present at least some promise of achieving via science some of the more radical promises that religion has traditionally offered—immortality, dramatic states of bliss, maybe even resurrection.

## We're Not Nerds and It's Not a Rapture

**Phil Bowermaster:** I hate it when people call The Singularity "the rapture for nerds." Okay, number one, we're *geeks*, not nerds. And number two, it's not a rapture, it's . . . something else, all right? Shut up.

# RAY KURZWEIL

Ray Kurzweil is the premier celebrity of both transhumanist and singularitarian culture. If he's speaking—as he often does—at one of the movement's signature gatherings (like Singularity Summit), he's the headliner. With his gentle, friendly mien and understated sense of humor, he's well liked and accessible to the least of the enthusiasts.

Yes, everybody likes Ray Kurzweil in these transhumanist circles. But in contrast to the impression you may get from media write-ups about the movement, hardly anybody "follows" (i.e. agrees with) him. That is to say that in books like *The Age of Intelligent Machines* and the seminal *The Singularity Is Near*, Ray has written a series of very explicit predictions about how change will proceed as the result of exponential change—for example he predicts that thousand dollar machines will be able to emulate human intelligence by 2020; and by 2045, an equally accessible AI will be billions of times more intelligent than all of humanity, at which time we will have achieved The Singularity. And most transhumanist and singularitarian types have either minor quibbles or major differences with these and other highly defined prophesies.

Nevertheless, Kurzweil is their leading cheerleader in the greater world, and he is admired and loved for his nerdy charisma and for bringing the message of radical technological change to at least a goodly chunk of the teeming masses.

## TECHNICAL CRED

Kurzweil's credibility as the leading cheerleader for transhumanist change is boosted by his successful career in tech development. He was the principal developer of the first omni-font optical character recognition, the first print-to-speech reading machine for the blind, the first CCD flat-bed scanner, the first text-to-speech synthesizer, the first music synthesizer capable of recreating the grand piano and other orchestral instruments, and the first commercially marketed large vocabulary speech recognition software. In late 2012, he was hired by Google to be their Director of Engineering.

## FILM AND TV

Aside from authoring bestselling books like *The Singularity Is Near*, Ray has been at the center of two films.

*The Singularity Is Near: A True Story about the Future*, which Ray co-directed with Anthony Waller, mixes fiction and nonfiction and revolves around the story of Ramona, a virtual alter ego that Ray created for himself way back in the 1990s. (The love for genderbending is fairly ubiquitous in transhuman culture, albeit it's mostly men wanting an alternative feminine self.)

*Transcendent Man*, a portrait of the man as well as his ideas, shows that his pursuit of The Singularity and the immortality he expects to result therefrom is largely inspired by the death of his father—a composer and musician—when Ray was only twenty-two. Ray even hopes to "bring back" his father, by digitally recreating his personality in an avatar based on information about him culled from his records and memories recorded from friends and family.

Kurzweil made his first television appearance on that wonderful '60s TV quiz show *I've Got a Secret* (an honor he shares with Frank Zappa and Colonel Sanders), where he showed off a piano piece that was composed by a computer he'd built. In recent years, he's been on *Charlie Rose*, *The Colbert Report*, *60 Minutes*, *Glenn Beck*, and innumerable other programs.

## WHOLE LOTTA PILLS

In 2007, Kurzweil claimed that he takes two hundred fifty nutritional supplement pills every day.

*Written with Surfdaddy Orca*

## Take The Singularity Easy, But Take It

**Ben Goertzel:** I often think of Ray's vision of the future as a "kinder, gentler Singularity," and this comes across clearly in the film *(Transcendent Man)*. Ray is portrayed as wanting The Singularity to help everyone, in the same way that he wanted the Kurzweil Reader to help blind people read. While acknowledging there are massive surprises to come, Ray—with his dry, soothing businessman's voice and demeanor—projects a calm confidence that all will be well even as legacy humans become obsolete. In his vision, we will all one day merge into a massive global cyborg-mind, yet still retain our capacity for individual experience and joy.

**Extropia DaSilva:** The aspects of Kurzweil's work that gain the most attention are his seemingly incredible claims regarding massively intelligent machines, uploading the mind into cyberspace, nanomachines that will reverse senescence, and so on. But I think his most interesting point was made during an interview with John Brockman:

> The kinds of scenarios I'm talking about twenty or thirty years from now are not being developed because there's one lab that's sitting there creating a human-level AI in a machine. They're happening because it's the inevitable end result of thousands of little steps. Each step is conservative, not radical, and makes perfect sense. Each one is just the next generation in some company's product.

# ROBOTICS

*(See also Artificial Intelligence, Cyborgs, Exoskeletons, Neurobotics, Sexbots, Warbots)*

## THE RISE OF THE ROBOTS: THIS TIME WE REALLY MEAN IT

The history of robots is a good lesson in speculation and technology. The concept of robots is ancient, but they entered wide public consciousness in the 1920s, and have been a staple of science fiction since then. By the 1950s there were worries about "automation" causing mass unemployment, but industrial robots only began working on manufacturing production lines in the 1960s. The Roomba robotic vacuum cleaner, arguably the first popular household robot, didn't arrive until 2002, and it hasn't replaced traditional vacuum cleaners.

So for generations robots were over-hyped, and unless you worked on an assembly line, you probably never saw one. That's now changing. Robots are flying in the air, walking across proving grounds, and swimming in the sea. Google has become a major player in robotics, experimenting with driverless cars and buying a bunch of robotics companies. Samsung wants to put a robot in every home by 2020, and while that may not mean Jetsons-style robot maids, all forms of robotics are rapidly advancing.

## CONTROLLED VS. AUTONOMOUS, ROBOTS VS. CYBORGS

These days robotics includes lots of things you might not think of as robots. Military unmanned aerial vehicles (UAVs) are essentially robots that fly with little human intervention, but often have human controllers to make some decisions. Robotics also includes some forms of human augmentation, such as telepresence robots, or powered exoskeletons that promise to help paraplegics walk and to make Tony Stark's Iron Man suit a reality. When not autonomous, robots are controlled via standard means such as computer screens and joysticks, but better brain-computer interfaces could allow them to be controlled by "thought helmets."

Researchers at North Carolina State University have added microchips and wireless transceivers to turn cockroaches into controllable "biobots" (though traditionally "cyborgs" is the term for robots that are partly biological). They hope to use these for things like searching for survivors in collapsed buildings.

Sharing the term, but not discussed here, are "bots" that are purely software. These are basically applications that run automated tasks over the Internet, for personal, commercial, or malicious purposes.

## MMMMM . . . BURGERBOTS

As is often the case with cutting-edge tech, the US military is leading the way in robotics, but there are also many peaceful robots. Ocean robots are doing environmental monitoring. Amazon wants to use GPS-guided drones to deliver small packages. Momentum Machines of San Francisco has developed a hamburger-making robot: just load it with ingredients and it makes and bags three hundred sixty gourmet burgers per hour. It only slices ingredients like tomatoes after the order is placed, for the freshest burger possible. The savings in labor will allow burgers made of gourmet ingredients to be sold at fast food prices. Even programmable robot toys have improved: sophisticated and affordable toy robots, often controlled via smartphone, are increasingly available.

## LEARNING FROM MOTHER NATURE

Robotics has benefited from biomimetics: using biology as models for the design and engineering of robots. This is especially important for microrobots, because propellors aren't the optimum propulsion for flying robots that weigh only grams. Years of research into insect flight has led to new varieties of insect drones that fly by flapping their wings, sometimes using piezoelectric actuators as artificial muscles. They are more maneuverable, more durable, and less likely to get entangled in obstacles than microdrones with props or rotors.

The Harvard Microrobotics Lab has RoboBees weighing just eighty milligrams and with a 1.2 inch wingspan, making them the smallest powered, heavier-than-air manmade objects to fly. While they include sensors for vision, the problems of a power source and a full onboard brain have yet to be solved, so each RoboBee has a tether for power and communication.

Larger and more capable microUAVs include Techject's six-inch, 5.5-gram robotic dragonfly, and AeroVironment's Nano Hummingbird. The size and weight of a large hummingbird, it can fly for up to eleven minutes while carrying a camera and communications system.

Not all miniature, biomimetic robots fly. The UC Berkeley Biomimetic Millisystems Lab has created the cardboard VelociRoACH, which can run up to seven miles per hour (twenty-six body lengths per second). It has six springy C-shaped legs, which spin up to fifteen times per second, and can carry four times its own weight.

## SWARM ROBOTICS

This is another way in which robotics is learning from nature. By enabling communication between them and having them follow simple rules, relatively simple robots can exhibit the decentralized, self-organized swarm intelligence seen in social insects.

As robots become more advanced, ethical questions arise. If a robot becomes sentient, does it have rights? Will they have feelings? And then there are the sexbots. Once they get realistic enough to escape the creepiness of the uncanny valley, they might have some social impact. But humanoid or not, mobile or not, deadly or peaceful, there will be lots of robots in your future.

**Ben Goertzel:** While not as well known as Sony's (now defunct) AIBO and Qrio or Honda's ASIMO, Samsung has its own line of humanoid looking robots, the Mahru bots. The latest, Mahru-Z, was created by engineers at KAIST building on earlier Samsung systems, and serves as a robot maid. You Bum-Jae, head of KAIST's cognitive robot center at the Korea Institute of

Science and Technology, brags: "It recognizes people, can turn on microwave ovens, washing machines and toasters, and also pick up sandwiches, cups, and whatever else it senses as objects." Mahru-Z is not yet ready for commercial dissemination, but it's clearly a serious step toward Korea's stated goal of "a robot in every home by 2020."

If, in fact, robots will be in every home, they'll need to know how to interact with humans emotionally as well as physically. Last summer I had the opportunity to learn something about Samsung's efforts in this regard, when Samsung engineer Hyun-Ryong Jung attended the Artificial General Intelligence Summer School that I hosted at Xiamen University in China. Among other topics, Jung told us about his work with Georgia Tech scientists implementing mathematical emotion models for humanoid robots. These models are inspired by cognitive science, but adapted to the particular requirements of digital beings and their interactions with humans. Among the interesting results found via studying the interactions of emotional robots and humans, Georgia Tech researchers demonstrated that younger people do better than older ones at recognizing emotions from robots' facial expressions.

## Robot Mannequins in Japan

**R.U. Sirius:** Those lucky mannequins. They got to have the look of botox injected bulimics way before it became truly fashionable, and now they get to become bots before us, too. And they even get to sell stuff to Japanese shoppers.

And speaking of *Minority Report* (well, maybe you were), a recent article in INVENTORSpot—titled "Japanese Robot Mannequins Move Shoppers to Buy" starts off: "Next time you check out that robot mannequin in a Tokyo department store window, don't be surprised if the mannequin is checking YOU out at the same time. 'Palette,' a robot mannequin, senses admiring shoppers and poses to please!"

On a slightly more sinister note, Flower Robotics is planning a more sophisticated program that will enable Palette to judge shoppers' age and sex and even recognize the logos on their shopping bags.

# SCIENCE FICTION

Transhumanism and singularitarianism are often criticized as being science fiction. Jason Pontin, currently the editor of *Technology Review,* has written, "I do think transhumanism is science fiction." No less a figure in technological progress than Craig Venter has called Kurzweil's projections "science fiction," as did the noted linguist and political activist Noam Chomsky.

It is true that most transhumanists will tell you that they've been inspired by some epic bits of science fiction (SF). (Not all: some don't read fiction at all!) Indeed, the very concept of a technological singularity in its contemporary sense was seeded primarily by SF writer Vernor Vinge with his 1986 novel *Marooned in Realtime,* which he then amplified in his 1992 novel *A Fire Upon the Deep,* and in a 1993 essay that appeared in *Whole Earth Review.*

But is this a bad thing? Some would argue that the future exists first in imagination. Many scientists and tech tinkerers have engaged— and continue to engage—in projects that were at least partly inspired by science fiction. In the '90s, we used to laugh about all the virtual reality projects (only partially successful) inspired by William Gibson's *Neuromancer,* since it was widely considered a dystopian vision. Among the innumerable inventions inspired by SF, we could include the helicopter (Jules Verne's novel *The Clipper of the Clouds),* the World Wide Web ("Dial F for Frankenstein" by Arthur C. Clarke), a mechanical hand called the waldo (named for a Robert Heinlein character), the taser (inspired by Tom Swift—also a major inspiration for original Apple computer creator Steve Wozniak), and Second Life (inspired by Neal Stephenson's cyberpunk classic *Snow Crash).*

## BIG WHOOP

Today, after the advent of face transplants, transgenic glow-in-the-dark salmon, and the first synthetic organism, SF is rumored to be losing some of its frisson with younger folks. They're living in a science *faction* world.

## SMARTER THAN US

Science fiction novels and short stories have been full of smarter than human systems, among the more interesting: "Evitable Aliens" by Isaac Asimov, *Childhood's End* by Arthur C. Clarke, and *Do Androids Dream of Electric Sheep?* by Philip K. Dick.

## IMMORTALITY HO!

Long-lived or immortal characters in SF trace back, at least, to Olaf Stapledon's 1935 novel *Odd John*, but the search for immortality in fiction goes all the way back to the *Iliad*. Mary Shelley, who famously fictionalized the creation of new life by raising the dead through technology in *Frankenstein,* also wrote a short story called "The Mortal Immortal" about a man tormented by his immortality. Twentieth century SF novels and short stories with an immortality theme worthy of mention include Brian W. Aldiss' "The Worm That Flies," Jack Vance's *To Live Forever*, and Roger Zelazny's *This Immortal*.

## HELP! I'VE FALLEN IN AND I CAN'T GET OUT—VIRTUAL WORLDS

Much literature has centered on virtual worlds. (Plato believed we were *in* one.) Aldous Huxley's *Brave New World* had the "feelies"—depicting virtual worlds as one of the many escapes and distractions for humans living in an utterly distracted and controlled condition. Vernor Vinge was among several who gave us humans trapped inside a simulation in his 1981 short story "True Names." The idea of uploading human minds onto computers was touched on in the 1950s in stories by Isaac Asimov and Arthur C. Clarke, and the topic has been explored more closely in works by Rudy Rucker and Greg Egan, among many others.

## CONTEMPORARY WRITERS

Among the current writers whose works are relevant to transhumanism, we would include Charles Stross, Cory Doctorow, the aforementioned Vinge, Rucker, Egan, Stephenson, Iain Banks, and Gregory Benford. Novice novelists who have emerged directly out of the transhumanist movement include Ramez Naam and Zoltan Istvan. And as noted by Jonathan Dotse of the AfroCyberPunk blog, science fiction is now influencing the developing world, too, so expect some science fiction writers to arise there. Everyone now lives in a science fiction world.

**Michael Moorcock (interviewed by Woody Evans):** The characters in [my 1981 novel] *The Dancers at the End of Time* are, in fact, the ultimate in transhumanism. They are immortal, pretty much omnipotent, and not unhappy. All such stories before were essentially dystopian, saying, "You can be immortal, without pain or hunger—but such conditions make you ultimately miserable." I wanted to write a story in which such people were actually pretty cheerful.

## Charles Stross

**Paul McEnery and R.U. Sirius:** Stripped to the high concept, the visions from Charlie Stross are prime geek comfort food. But don't be fooled. Stross' stories turn on you, changing up into a vicious scrutiny of raw power and the information economy.

The "God" of *Singularity Sky* is really just an Artificial Intelligence, manipulating us all for the sole purpose of beating the alien competition. The Merchant Princes (from a series of novels by Stross) are just as rapacious as anything on Wall Street, and a downstream parallel universe is just another market to exploit. *The Atrocity Archives* gives us a gutpunch full of paranoia—on the far side of hacking and counterhacking lurks an unspeakable chaos. And for all our engineering genius, *Accelerando's* paradise is won at the cost of planetary destruction, with humanity at a dead end as our future heads off into the stars without us.

In *Halting State*, Stross savages the fantasy worlds we escape into for fun and profit and invites us to peek underneath the surfaces as our chattering gadgets dress up reality with virtual sword-and-sorcery games, all underwritten by oh-so-creative financial instruments.

All of Stross's highly connective pipe-dream superstructures are wide open to the one geopolitical prick that will pop them all like the balloon animals they are. Be warned. Take care of the bottom line, or your second life will cost you the life that counts.

It's no surprise that Stross is a highly controversial figure within transhumanist circles—loved by some for his dense and high-concept takes on themes dear to the movement, loathed by others for what they see as a facile treatment of both ideas and characters. But one thing is certain—Mr. Stross is one SF writer that pays close attention to the entire plethora of post-humanizing changes that are coming on fast. As a satirist, he might be characterized as our Vonnegut, lampooning memetic subcultures that most people don't even know exist.

# SENS RESEARCH
# FOUNDATION

*(See also Aubrey de Grey, Longevity/Immortality, The Methuselarity)*

## REJUVENATION BIOTECHNOLOGY

A leader in rejuvenation research, the nonprofit SENS Research Foundation was founded in 2009 by Michael Kope, Aubrey de Grey, Jeff Hall, Sarah Marr, and Kevin Perrott. Their name stands for Strategies for Engineered Negligible Senescence, and is meant to highlight their approach to anti-aging research, which is "the repair of living cells and extracellular material in situ."

The most common anti-aging strategies either focus on altering metabolism to prevent the accumulation of aging damage, or attempt to minimize the effect of this damage. In contrast, the SRF wants to selectively target and repair the seven major kinds of aging damage: cell loss and tissue atrophy, cancerous cells, mitochondrial mutations, death-resistant cells, extracellular crosslinks, extracellular junk, and intracellular aggregates.

The SRF does research in-house at their center in Mountain View, as well as funding research elsewhere. They have a variety of outreach and education programs, including conferences, workshops, a student internship program, and are developing open online coursework.

# SEXBOTS

*(See also Robotics)*

It's been something of a cliché in tech culture that new media tech-nologies get popularized among consumers as a tool for pornography. Among the techno-toys and breakthroughs that found their earliest markets largely as a tool for enjoying porn, we would include VHS, interactive TV, Blu-ray, online payment systems, VoIP (Voice Over Internet Protocol), and broadband.

This has not been the case with robots. Still, one day, someone will build an attractive, soft, fleshy-feeling bot, and then all bets are off.

## HARD FOR WOMEN? (NYUK-NYUK-NYUK)

In a rare reversal of fortune, the common electric vibrator has long put women at the forefront of sex machines, although vibrators can scarcely be mistaken for sexbots. Still, the mechanics of sex may favor the initial popularization of male robo-servants (hard being more easily built than soft and fleshy).

Given the marketability of sexual pleasure, sexbots are inevitable. Will they destroy intimacy? Or has porn done that already?

**Hank Pellissier:** The sexbots are coming, and we will cum with them.

Remember the most convulsive, brain-ripping climax you ever had? The one that left you with "I could die happy now" satiety? Sexbots will electrocute our flesh with climaxes twice as gigantic because they'll be more desirable, patient, eager, and altruistic than their meat-bag competition, plus they'll be uploaded with supreme sex skills from millennia of erotic manuals, archives, and academic experiments, and their anatomy will feature sexplo-sive devices.

Sexbots will heighten our ecstasy until we have frothy, shrieking, bug-eyed, amnesia-inducing orgasms. They'll offer us split-tongued cunnilingus, open-throat fellatio, deliriously gentle kissing, transcendent nipple tweaking, g-spot massage, and prostate milking dexterity, plus two thousand varieties of coital rhythm with scented lubes—this will all be ours when the sexbots arrive.

When will they get here? Henrik Christensen, founder of the European Robotics Research Network, predicted we'd be boinkin' 'bots by 2011. Whoops! Dr. David Levy, author of the recent book, *Love and Sex with Robots*, believes by 2050 these robots will be nearly indistinguishable from humans.

## Eww . . . Gross!

**Hank Pellissier:** Are sexbots icky? Are humans pathetic if we don't just mate with each other? Truth is, we're already mostly "solo" when it comes to orgasms. "Masturbation," noted Hungarian-born psychiatrist Thomas Szasz, "is the primary sexual activity . . . in the 19th century it was a disease, in the 20th it's a cure." Sure, we generally prefer sex with live partners, but the desired one is often unavailable or inadequate. Sexbots will never have headaches, fatigue, impotence, premature ejaculation, pubic lice, disinterest, menstrual blood, jock strap itch, yeast infections, genital warts, AIDS/HIV, herpes, silly expectations, or inhibiting phobias. Sexbots will never stalk us, rape us, diss us on their blog, weep when we dump them, or tell their friends we were boring in bed. Sexbots will always climax when we climax if we press that little button on their butt.

## Roxxxy the Female Sexbot

**R.U. Sirius:** Roxxxy, a female sexbot, wowed crowds at 2010's AVN Adult Entertainment Expo in Las Vegas. According to her designer, Roxxxy "has a full C cup and is ready for action."

So it starts, of course, with a boy toy, which somewhat offends our sense of gender equality.

# SIMULATION THEORY

*(See also Artificial Life)*

## IS REALITY REALLY REAL?

What if the universe is actually some sort of simulation? This may sound like the idle speculation of dorm-room stoners and fans of Philip K. Dick and *The Matrix*, but it has more modern philosophical and even scientific support than you might think.

The idea that reality is an illusion, something beyond human comprehension, has a long philosophical pedigree. It dates back at least as far as the pre-Socratic philosophers, and appears (in various forms) in ancient Hindu and Buddhist teachings. Descartes took it a step further, introducing the skeptical hypothesis to modern Western philosophy by postulating an omnipotent "evil demon" devoted to presenting a complete illusion of the external world.

## IS GOD A COMPUTER PROGRAMMER?

The advent of computers, with their ability to simulate complex systems, made the concept more plausible. While the laws of physics appear continuous, quantum physics tells us that everything in space and time is quantized: granular at the smallest possible level. (So Democritus was partly right about atomic theory, but the indivisible bits he called atoms turned out to be much, much smaller than he could have imagined.) Since everything is made of quanta, the way a digital image is made of pixels or a software program is made of bits, in theory a large enough computer could calculate everything we experience, and create a simulation of the entire universe.

Hans Moravec and Nick Bostrom have both theorized about this, and Bostrom and others even consider it *likely* that our universe is a simulation run by other intelligences, possibly a future posthuman civilization. So if we're not "real," do we still have rights as humans? If so, wouldn't it then follow that if we created artificial general intelligences, or uploaded human minds to computers, that they would deserve the same rights as we do?

And it's also possible that the intelligences who created us are themselves living in a simulation . . .

## HOW WOULD WE EVER KNOW?

So is there a way for us, the simulated, to detect the simulators? Maybe. We might notice physical constants like the speed of light inexplicably changing, due to some sort of simulation software upgrade. And indeed, in 1999 some astronomers thought they'd discovered that the fine-structure constant is fractionally larger today than it was ten billion years ago. However, this was not confirmed (or maybe the simulators fixed the glitch before it could be confirmed).

Supercomputers are now using lattice quantum chromodynamics (lattice QCD) to model and test the universe at the most fundamental level, in areas just a few femtometers across. However, there are still resource constraints that reveal that these are simply simulations. Using lattices is still a shortcut, and the spacing of a lattice imposes an upper limit on particle energy, as well as a lower limit on the possible size of anything.

This has led to the thought that, if we are in a simulation, there would be an upper limit on the energy of cosmic rays . . . and there is: the Greisen-Zatsepin-Kuzmin (GZK) limit. Interaction with the cosmic microwave background radiation is thought to explain this cutoff, but if we are in a simulation, cosmic rays could reveal the existence and orientation of the lattice by preferentially traveling along the axes, meaning that we would not see cosmic rays coming from all directions equally. If we detect that, we'd be seeing a shortcut our simulators are using.

That is, unless this universal lattice is constructed in a different way than thought, or has a much smaller spacing, or these advanced minds have figured out a way to construct a simulation without using lattices at all. Then we wouldn't see any variation from what current physics expects, and simulation theory would remain unproven.

Simulation theory could also be falsified: if something uncomputable is discovered, it would prove that reality is doing something no computer can possibly do, so we couldn't be inside of one.

In any case, we'd like to know the truth, so we hope that our readers will get right to work on settling this question.

## Living in a Simulation in a Matrioshka Brain

**Surfdaddy Orca:** "Dyson spheres," a concept from the theoretical physicist Freeman Dyson, would consist of a system of orbiting solar power satellites meant to surround a star and capture most or all of its energy output. Computer scientist Robert Bradbury took the concept of Dyson spheres and proposed nesting them inside one another like Russian matrioshka or babushka dolls using nanoscale computers—creating essentially a giant brain. Nick Bostrom, Ray Kurzweil, and others speculate that an advanced civilization may have already created such a brain, and that we humans are simply simulations running inside it.

## Who Cares?

**Seth Arthur Weisburg:** Bostrom's Simulation Argument is mostly irrelevant—it doesn't address the core issue. If our universe was created by some alien programmer as part of a third-grade class project, that means something—if there is some way for us to observe or interact with that programmer, and learn new things about our universe or his. But if there's no way to observe the putative "meta level" in which the creation of our universe as a simulation was performed, then the hypothesis of the universe as a simulation has no real significance. It doesn't matter if we call the particles and fields in terms of which science models our world "simulation" or "physical reality." What matters is whether the observations we make as individuals—from which we infer the validity of scientific models—are existent in an intersubjective sense or not . . . Whether our world is a "simulation" or not is a chimerical question.

# THE SINGULARITY

For some futurephiles, The Singularity is the Big Kahuna. Science fiction writer and futurist Arthur C. Clarke famously said, "A sufficiently advanced technology is indistinguishable from magic." And The Singularity would be, according to some, the ultimate magic bullet—no more death, no more scarcity, no more suffering, no more ignorance and stupidity—just biological life and machines achieving unimaginable intelligence with the problems that plague humanity solved and new worlds to conquer. According to some, including singularitarian main man Ray Kurzweil, we will even seed the galaxy with the intelligence we create.

## THE TECHNOLOGICAL SINGULARITY

As you know by now, the generally accepted notion of The Singularity (more correctly called The Technological Singularity to distinguish it from singularities that occur in a whole bunch of different fields including mathematics and geometry) is that it's the time when our artificial intelligences become substantially smarter than human beings. Most people who declare themselves singularitarians expect to see this occur within their lifetimes, even perhaps their unextended lifetimes. In other words, some time in this century. Ray Kurzweil predicts The Singularity in 2045.

The contemporary notion of The Singularity got rolling with acclaimed SF writer Vernor Vinge, whose 1981 novella *True Names* pictured a society on the verge of this event. In a 1993 essay, "The Coming Technological Singularity," Vinge embraced his fictional vision as future fact, writing that "within thirty years, we will have the technological means to create superhuman intelligence. Shortly after, the human era will be ended."

The release of Ray Kurzweil's *The Singularity Is Near: When Humans Transcend Biology*—which was near the top of *The New York Times* Best Seller list for much of 2005 and 2006—probably did more than anything else to raise awareness of The Singularity.

## HARD TAKEOFF VS. SOFT TAKEOFF

A popular discourse among singularitarians revolves around the question of whether we will have a hard or soft takeoff into The Singularity.

A hard takeoff revolves around the notion that exponential growth in computing power could cause AIs to suddenly become vastly superior to humans in a very short time—months, weeks, days, or possibly even moments.

In exponential multiplication 8 x 8 = 64; then 64 x 64 = 4,096. Now, let's say that you're up around ten million and your AI implant is solving most of your problems for you. In fact, it's solving its own problem of continuing the exponential growth in processing power so fast that it hits its next exponential in a matter of days. So to multiply 10,000,000 x 10,000,000 in CPU power leaves you with 100,000,000,000,000 (100 trillion). Suddenly your AI is about five to ten million times smarter than you. Everything it can say and do that might be useful is way beyond your comprehension. You just hope that it likes having a pet.

A slow takeoff would involve a slower improvement. This could occur for a variety of reasons, among them the fact that processing power doesn't necessarily translate that quickly or easily into the equivalent of the sorts of learning through experience that humans are capable of. The slow takeoff could also take place by human intention. Humans would consciously remain in control of the rate of change to insure that the AI is friendly to human concerns, or perhaps to the private concerns of some controlling group.

## DIFFERENT SCHOOLS OF THOUGHT

According to Eliezer Yudkowsky, who started the movement to ensure we develop "friendly AI," there are three schools of thought about the nature of The Singularity.

1. *Accelerating Change:* This is the predominant discourse, thanks perhaps to Kurzweil's advocacy. AI power will grow exponentially leading to the upcoming Singularity.

2. *Event Horizon:* Emphasizes the point at which human beings are no longer the most intelligent beings on the planet and how in-comprehensible and strange the future is for us beyond that point. Not surprisingly, the emphasis on strangeness comes from the science fiction writer Vernor Vinge. Event Horizon doesn't seem to contradict Accelerating Change, but the different emphasis leads to different types of discussions.

3. *Intelligence Explosion:* Emphasizes human intelligence, placing humans at the center of technological evolution. The Intelligence Explosion sees us creating smarter-than-human intelligence inter-faces but still driving them with our own intention.

## POP GOES THE SINGULARITY

Transhumanism may be envious of The Singularity's brand recognition. If I approach educated strangers and mention transhumanism, the odds are about ten to one that they'll look at me quizzically and say, "What's that?" But if I mention The Singularity, there's probably about a one in three chance they'll at least have heard of it. So The Singularity—a highly speculative and viscerally rather scary notion—has more memetic juice than mere human enhancement, something that is already occurring. Such is the nature of what draws people's attention.

## Outsourcing Cognition Is a Sign Toward The Singularity

**Vernor Vinge (interviewed by Doug Wolens):** Humans may not be best characterized as the tool-creating animal, but as the only animal that has figured out how to outsource its cognition—how to spread its cognitive abilities into the outside world. We've been doing that for a little while—ten thousand years. Reading and writing is outsourcing of memory. So we have a process going on here, and you can watch to see whether it's ongoing. So, for instance, in the next ten years, if you notice more and more substitution for using fragments of human cognition in the outside world—if human occupational responsibility becomes more and more automated in areas involving judgment that haven't yet been automated—then what you're seeing is rather like a rising tide of this cognitive outsourcing. That would actually be a very powerful symptom [of the coming singularity].

## Singularity When?

**Vernor Vinge (interviewed by Doug Wolens):** I'd personally be surprised if it hadn't happened by 2030. That doesn't mean that terrible things won't happen instead, but I think it is the most likely non-catastrophic event in the near future.

## Scary But Promising

**Vernor Vinge (interviewed by Doug Wolens):** You are contemplating something that can surpass the most competitively effective feature humans have—intelligence. So it's entirely natural that there would be some real uneasiness about this. The nearest analogy in the history of the Earth is probably the rise of humans within the animal kingdom. There are some things about that which might not be good for humans.

On the other hand, I think this points toward something larger. Thinking about the possibility of creating or becoming something of superhuman intelligence is an example of an optimism that is so far-reaching that it forces one to look carefully at what one has wanted. In other words, humans have been striving to make their lives better for a very long time. And it is very unsettling to realize that we may be entering an era where questions like "what is the meaning of life?" will be practical engineering questions. I think it could be kind of healthy, if we look at the things we really want, and look at what it would mean if we could get them.

## The Singularity Is Bollocks: A Skeptical View

**Ramez Naam:** Part of the justification for the term "The Singularity" is the idea of an event horizon in the future, a point beyond which we are simply not able to predict anything about the world. We are, in theory, approaching this point rapidly. It may be just thirty or forty years away, a looming phase transition that will lead to a world we simply cannot understand. Well, bollocks to that, I say.

The reality is that our ability to understand the future—and the length that our "future headlights" can peer, if you will—is at an all-time high. The event horizon, if there is one, is receding all the time.

How can I say this? Wouldn't the creation of AI, or the uploading of humans into computers, or the end of aging, or the creation of self-replicating molecular nanotechnology, or some other advance change the world so fundamentally that we couldn't understand it?

Nope.

All of those phenomena are still governed by the laws of physics. We can describe and model them through the tools of economics, game theory, evolutionary theory, and information theory. It may be that at some point humans or our descendants will have transformed the entire solar system into a living information processing entity—a matrioshka brain. We may have even done the same with the other hundred billion stars in our galaxy, or perhaps even spread to other galaxies.

Surely that is a scale beyond our ability to understand? Not particularly. I can use math to describe to you the limits on such an object, how much computing it would be able to do for the lifetime of the star it surrounded. I can describe the limit on the computing done by networks of multiple matrioshka brains by coming back to physics, and pointing out that there is a guaranteed latency in communication between stars, determined by the speed of light. I can turn to game theory and evolutionary theory to tell you that there will most likely be competition between different information patterns within such a computing entity, as its resources (however vast) are finite, and I can describe to you some of the dynamics of that competition and the existence of evolution, co-evolution, parasites, symbiotes, and other patterns we know exist.

# SINGULARITY UNIVERSITY

Formed in 2009 by XPRIZE founder Peter Diamandis, along with Ray Kurzweil, Singularity University (SU) announced itself as a place to "educate, inspire, and empower leaders to apply exponential technologies to address humanity's grand challenges." Founded as the result of a meeting hosted by NASA's Ames Research Center at Moffett Field, California, and sponsored by Google, Nokia, and Genentech (among others), SU has placed the singularitarian notion in a distinctly establishment framework.

## LOTS OF MONEY. NO CREDIT.

Singularity University is not, as of yet, an accredited university, and it costs a good chunk of change to get in. Its Graduate Studies Program, ten weeks over the summer, is limited to eighty students and costs $29,500. Students in the Graduate Program are challenged to come up with technologies that can "positively impact at least one billion people within ten years." That'd be tough to measure, and you'd have to wait ten years to find out if you "passed." What if The Singularity comes first?

The point, of course, isn't the conventional college degree, but the seeding of a lot of projects and careers. SU puts its students in touch with a lot of powerful people and groups who can give them a head start on achieving their notions.

## A SINGULARITARIAN TRAINING CAMP?

SU isn't really about The Singularity, per se. It's not a training ground for Singularity believers—although that would be fascinating, in a reminds-me-of-Scientology sort of way. In fact, most who have taught courses there are fairly widely known figures in the tech world and not particularly of the singularitarian persuasion. These include World Wide Web inventor Vint Cerf, Tim Ferriss, Larry Brilliant, Google AI researcher Peter Norvig, and Ethernet co-inventor Robert Metcalfe.

## ONGOING

SU seems to have succeeded in becoming an ongoing institution and has converted from a nonprofit into a for-profit corporation. I guess we'll have to wait until 2019 to find out if anybody has positively impacted a billion people, something that hasn't been done since the invention of ice cream. (I just made that up.)

### Problem Solving

**Peter Diamandis (interviewed by Alex Lightman and R.U. Sirius):** We're going to be asking the students to focus these tools—these extraordinarily powerful tools coming out of these exponentially growing fields—on the world's biggest problems. We have these large, global, intractable problems: pandemics, hunger, energy, whatever it might be. And the only way we'll be able to handle them is by wisely using the power of these exponentially growing technologies.

### Singularity? Just a Name

**Peter Diamandis (interviewed by Alex Lightman and R.U. Sirius):** To be clear, the university is not about The Singularity. It's about the exponentially growing technologies and their effect on humanity. Now, one of the potential outcomes can be what has been referred to as The Singularity.

You know, we toyed with other terms, like Convergence University.

# SOUSVEILLANCE

*(See also Quadcopters)*

By now, everybody knows that "they" are watching "us." But what if "we" are watching "them"? And what if everybody is watching everybody? We have met Little Brother and he is us.

## MANN UP

Steve Mann, who was wearing what he called "computerized eyeware" around the time Larry Page and Sergey Brin would've been entering kindergarten, came up with the term "sousveillance" to describe the growing number of individuals who are equipped to record events in public on a moment's notice, or as they occur.

Sousveillance, then, is people's surveillance. With regular folks wandering around wearing (or even just carrying) recording equipment (i.e. anyone with a smartphone), and with this data, in some cases, viewable in realtime over the Internet, sousveillance is viewed as a great equalizer. Already, life has been made more difficult for those who would abuse their authority, with dozens of cases of policemen, TSA officers, and the like caught in acts of brutality and excess.

As Mann puts it, "Sousveillance—the inverse of surveillance—is the general activity of an individual capturing a first-person recording of an activity from his or her own perspective as a participant in the activity. Rather than watching 'from above,' the French 'sous' means 'under' or 'from below.'"

## SO HOW ARE WE ENHANCED BY SOUSVEILLANCE?

For one, our memories fail. And even if our recollections of events are accurate, we can't show someone else what we saw. Now, as the technology becomes closer to us—becomes something some of us live with during a goodly portion of our lives—we become creatures who record what we see and can share it, instantly, with potentially a few billion people.

Someday this shit may go behind the eyes. Hopefully we'll have a mechanism for turning it off when we want to.

## Steve Mann's Predictions

- Prediction #1: *Sousveillance will arise naturally as we replace failing memory with wearable technology.* Just as buildings are automated, and this automation leads naturally to—and is integrated with—surveillance, putting "intelligence" (wearable computing) onto people will give rise to widespread sousveillance.

- Prediction #2: To the extent that communications prosthetics result in decreased overall risk—or risk perceptions—sousveillance will become the expected social norm. In much the same way that good risk management practice includes surveillance of buildings, the same will be true of people's protection of their own personal spaces.

- Prediction #3: *Surveillance and oversight will become less necessary when there is sousveillance* due to the increased availability and miniaturization of cameras and wearable computers. If everyone remembered (recorded) everything, there would be little need for surveillance.

- Prediction #4: *A sousveillant society will be a society with less corruption.* Where there is widespread sousveillance, corruption will be either impossible or at least very difficult. While surveillance locks the basement doors, sousveillance locks all the doors. A security program without sousveillance is not real security.

# SPACE COLONIZATION

In many ways, the transhumanist meme harks back to a movement for space colonization and the L5 Society that formed around that cause during the 1970s. Timothy Leary was enthusiastically involved in L5, as was the "father of nanotechnology," a young K. Eric Drexler. Early lights in the extropian movement met at L5 events. Among them were Keith Henson, a founding extropian, and Hans Moravec, the robotics professor who was kind of the go-to guy for extreme predictions about posthuman AI before Ray Kurzweil.

In September 1974, physicist and Princeton professor Gerard K. O'Neill published a paper in *Physics Today* suggesting that artificial living environments (i.e. space colonies) could be built at Lagrange points—locations in Earth's orbit where a habitat could theoretically remain stable. One of these stable points was called L5, and the term was adopted by the space colonies movement. Aside from future transhumanists, the movement received enthusiastic support from California Governor (then and now) Jerry Brown, which is how he acquired the moniker Governor Moonbeam.

Space colonies didn't happen (too expensive) and many of the enthusiasts followed K. Eric Drexler, who believed that nanotechnology was the answer to the problem of getting off Earth. The idea was that control over the molecular structure of matter would produce the material wealth, clean energy, and powerful yet light spacecraft that would allow us to get out there and start building the High Orbital Mini Earths (as Leary liked to call them).

Well, the sort of nanotech that Drexler has in mind is still tantalizingly on the horizon, but there have been some new movements in the direction of up-and-out in recent years. The Russian company

Orbital Technologies is promising to open a space hotel two hundred miles above Earth by 2016. And Richard Branson's Virgin Galactic also promises a space hotel (as well as a trip to the Moon) really soon.

## FLOOR 7,827,000 PLEASE

There is an evolving movement advocating for a space elevator as a means of reaching orbit economically. The idea is to build a tether between Earth and orbit. The materials required need to be both incredibly strong and very light. Recent developments in carbon nanotubes and boron nitride nanotubes suggest that a sufficiently strong and light material may be within reach. A space elevator could bring lots of solar energy to Earth, enable zero-G manufacturing, and eventually make space colonies a reality.

**Peter Diamandis (interviewed by Alex Lightman and R.U. Sirius):** I had the pleasure of flying Stephen Hawking into zero-G. I asked Hawking why he was doing this. And he answered—before the media at the press conference—that he believed that if the human race does not evolve into space, we don't have a future. Because there are so many problems—with asteroids, pandemics, and war—that we, effectively, have to back up the biosphere. So opening the space frontier is critical for the purpose of backing up the biosphere, and for getting access to the resources needed for the continual growth of humanity.

**Howard Bloom (interviewed by R.U. Sirius):** We have to bring space to life by bringing life to space. Economies go through mega-crashes roughly once every seventy years. What lifts economies from those massive falls? New frontiers, new resources, new technologies, new ways to turn toxic wastes into energy, and new techniques that turn garbage into gold.

We evolved as humans by turning hostile environments to our advantage. We evolved as humans by finding ways to live on the edges of Ice Age glaciers and on the fringes of deserts. We evolved by outfoxing sixty freezes and eighteen periods of massive global warming. Today we have a new frontier whose potential is larger than any hostile horizon we've ever conquered before. It's a new frontier not just for human beings but also for entire ecosystems—for meshes of living beings like bacteria and algae to trees, cats, and puppies. It's a massive niche waiting to be greened. And we are the only beings on this planet that can reach it. We are the only ones who can green it. That vast new landscape hangs above our head.

# STEAL THIS SINGULARITY

A project by your lead author R.U. Sirius, begun in 2012.

1. The notion that the current and future extreme technological society should not be dominated by big capital, authoritarian states, or the combination thereof. Also related, a play on the title of a book by 1960s counterculture radical Abbie Hoffman.

2. The notion that in our robotized future, human beings shouldn't behave robotically. The well-rounded posthuman—if any—should be able to wail like a banshee, dance like James Brown, party like Dionysus, revolt like Joan of Arc, and illuminate the irrational like Salvador Dalí.

3. The title for a website in which R.U. Sirius says and does as he pleases.

Steal This Singularity has almost nothing to do with the notion that we will develop artificial intelligences that are smarter than us or that if such a thing were to happen it would be a "singularity." I just like the name.

# STEM CELLS

*(See also Body Sculpting, In Vitro Meat)*

Stem cells are undifferentiated cells, found in all multicellular organisms. They are a sort of raw material your body uses to create more specialized cells: skin, muscle, bone, blood, heart, and so on. Their many uses have caused a revolution in medicine (and some controversy) in recent decades. They can be grouped into three basic types.

## EMBRYONIC, ADULT, AND INDUCED PLURIPOTENT

Embryonic stem cells develop and differentiate to form any type of human tissue, and so they are called pluripotent. They are harvested from embryos, can be grown indefinitely in a lab, and are at the heart of the bioethics controversy we'll touch on below.

Adult stem cells (sometimes called tissue stem cells, or tissue-specific stem cells) are the sort we all have in various subtypes, maintaining and repairing our tissues throughout our lives. They are not pluripotent, but specialized. For example, bone marrow stem cells continually become only red or white blood cells, not anything else. These are usually harvested from blood (including umbilical cord blood from newborns), fat, or bone marrow, and can't yet be grown indefinitely in a lab.

Induced pluripotent stem cells are a 2006 discovery by Shinya Yamanaka, which won him (along with John B. Gurdon) the 2012 Nobel Prize in Physiology or Medicine. These are mature, specialized adult cells that have been reprogrammed to regain their pluripotency, and so act like embryonic stem cells.

## AVOIDING ETHICAL ISSUES

The stem cell controversy centers on the use of human embryos. Destroying three- to five-day-old embryos for their stem cells strikes some people as immoral. Thankfully, new techniques for using adult stem cells, amniotic stem cells, and especially induced pluripotent stem cells have largely rendered this issue moot. Another reason for this shift is that embryonic stem cells often turned out to be more dangerous than hoped, causing side effects such as cancer.

## TYPES AND METHODS OF STEM CELL THERAPY (SCT)

Autologous SCT uses stem cells from your own body, while allogeneic or heterologous SCT uses stem cells from someone else. Both types have advantages and disadvantages.

Administering SCT can be done by injecting the cells directly into the blood or tissue, but the cells may not go exactly where they are needed. One new method is to use a modified inkjet printer to "print" an optimal pattern with precisely sized droplets, and build up layers of tissue that grow together into bone, muscle, or whatever is needed, which is then implanted in the body.

## USES OF SCT

SCT is a major tool for regenerative medicine, and has been since the development of the oldest form of adult stem cell therapy: bone marrow transplants to treat leukemia. In recent years, the field has exploded, so there are now too many uses to list. Let's just say SCT includes present and future treatments for cancer, Parkinson's disease, brain and spinal cord injuries, ALS, MS, muscular dystrophy, Crohn's disease, autism, HIV, epilepsy, cerebral palsy, damaged hearts, blindness, deafness, baldness, arthritis, diabetes, and wound healing. Some previously untreatable conditions have been entirely cured.

## REGENERATION AND BEYOND

Stem cells are also used to create synthetic blood, for testing new drugs, and may eventually be used to grow entire replacement organs or even limbs. They could enable transgender surgery with full reproductive capabilities. Stem cells are now used to reconstruct breasts damaged or removed by cancer treatments, and even healthy breasts can be enlarged (within limits) with this method. If that isn't enough to thrill many women and men and the entire porn industry, researchers have also used stem cells to enlarge rabbit penises. It's difficult to imagine that commercial possibility being ignored.

## ONE STEM CELL BURGER, TO GO

Physiologist Mark Post of Maastricht University, worried about the environmental impacts of meat production, reached a breakthrough in 2013 by cooking and eating a hamburger patty made from lab-cultured stem cells. The verdict: good taste and excellent mouth feel, but without fat cells it lacked some juiciness. Like much else related to stem cells, it still needs work.

# SYNTHETIC BIOLOGY

*(See also Biohacking, Cloning, Genomics. For digital simulations of life, see Artificial Life.)*

Just as chemistry advanced from the study of natural chemicals to designing and creating new ones, biology has gone from studying life to engineering molecules, cells, tissues, and entire organisms.

In a way, we've been creating new organisms for millennia, via selective breeding. Nothing quite like your poodle or your popcorn existed in ancient times. Today, the great strides made in manipulating DNA are enabling extreme crossbreeding and more precise and extensive modifications. Much of this work is still just emerging from the labs, but engineered microorganisms can produce biofuels, pharmaceuticals, industrial enzymes, and "greener" versions of petrochemicals like artificial rubber and acrylic. They can detect and destroy chemical or biological agents, remediate pollution, and create useful chemicals from agricultural waste. Synthetic biology is even enabling the creation of living things nearly "from scratch." It looks like another industrial revolution is in the making.

## MEET THE SPIDER-GOATS

Spider silk is an amazing substance, five times stronger than steel and more elastic than Kevlar, but it's hard to get much of it. You can't farm spiders because they eat each other. Both Nexia Biotechnologies of Canada (now bankrupt) and Professor Randy Lewis at Utah State University have worked around this problem. A gene is taken from an orb-weaver spider, added to a goat egg, and the egg is implanted in a mother goat. There are now dozens of spider-goats producing spider-silk protein in their milk. Lewis has done the same thing with silkworms, enabling them to create fibers twice as strong and twice as elastic as normal silkworm silk.

With enough spider silk, you could make improved scaffolding for artificial skin, better sutures, artificial ligaments, better bullet-proof fabric, and tougher and more resilient car fenders and suspension bridges.

## ANYONE CAN PLAY

Much of the activity in this field operates along the lines of open source software, with groups like the Registry of Standard Biological Parts, the OpenWetWare community, and the BioBricks Foundation promoting cooperation, standardization, and best practices. For more, see Biohacking.

## SYNTHETIC BIOLOGY: THREAT OR MENACE?

Of course there are ethical and safety concerns, both legitimate and exaggerated. Military types worry about terrorist biohackers creating bioweapons. Animal rights types are unhappy with what they see as exploitation of living creatures. Many people are frightened of GMOs, and "bioconservatives" don't much like anyone tampering with nature at all.

Thankfully, less panicky voices like the BioBrick Foundation are working to promote synthetic biology while avoiding the feared pitfalls. Synthetic biology is too promising, both scientifically and commercially, to be ignored or suppressed.

## DID VENTER MAKE SYNTHETIC BIO?

On May 20, 2010, Craig Venter and his team at Venter Institute announced that they had created synthetic genes and implanted them into living bacteria, which some might call the first actual creation of synthetic life. Their press release referred to it as "the first self-replicating cell controlled only by the synthetic genome." However, many skeptics argued that Venter was overhyping the move and that it was not a major development toward synthetic biology. In an *h+* article, research scientist and science journalist Dr. Alan Goldstein—who has published a list of criteria for identifying artificial life forms, said, "While technically impressive, the achievement of Venter's group is basically a linear extension of previous work in the growing field of Synthetic Biology. In fact, Genetic Engineers have been inserting large fragments of synthetic DNA and even synthetic 'minichromosomes' into cells for decades." Writing in *The Huffington Post*, Athena Andreadis, an Associate Professor of Cell Biology, said, "The Venter work is not a discovery, let alone a paradigm shift. It's a technological advance and even then not of technique but only of scale."

It looks like we'll have to wait a few more years before officially panicking over "man playing God by creating life."

# TELOMERES

## THE FOUNTAIN OF YOUTH . . . MAYBE

Telomeres are sections of repeated DNA at the ends of your chromosomes, protecting them from rearranging or fusing with other chromosomes, neither of which you want to happen. Every time a cell divides, the chromosomes replicate and the telomeres get shorter. Shorter telomeres are linked to cancer, heart disease, dementia, and other age-related diseases. Eventually, the cell reaches the Hayflick limit, and enters senescence.

However, your cells also produce an enzyme called telomerase, which replenishes telomeres, and that's where it gets interesting. Experiments with mice have produced spectacular results. Mice engineered to lack telomerase age much faster than normal mice, suffer from diseases like osteoporosis and diabetes, and die young. When their telomerase was turned on, the effects of aging were dramatically reversed: they got stronger, smarter, regrew their hair, and became fertile again.

Aha, you may think, just get me more telomerase! Alas, it's not that simple.

## WHY IT'S NOT SIMPLE

Eating or injecting telomerase won't help. The only known ways to get more involve drugs, gene therapy, or hibernation to induce cells to produce it on their own. And even those aren't sure things, because the relationship between telomeres and aging is complex. We know that some animals produce enough telomerase so that their telomeres don't shorten, and yet they still age and die. Some species have long telomeres and yet age relatively quickly, while others with shorter telomeres age more slowly.

Studies of several long-lived seabird species have also added to the confusion. The telomeres of Leach's storm petrel seem to lengthen with chronological age. The telomeres of the great frigatebird do decrease with age, but the rate of decrease varies between individual birds. And it's possible that telomerase activation might have negative side effects, such as encouraging tumor growth. Malignant tumors produce plenty of telomerase, which allows them to grow indefinitely until the organism dies.

Still, telomerase therapy might be a key component of life extension in the future.

**Rodney Shackelford:** Although much research is needed on the basic molecular functions of telomerase, it appears that a few relatively small genetic alterations in the mammalian genome and protein expression patterns, including increased telomerase expression, can result in a significantly longer lifespan and a reduction in age-associated diseases. Thus, it's very likely that telomerase will be a major target for genetic alterations designed to increase the human lifespan, remaining a very active area in anti-aging research.

## Rejuvenation

**Rodney Shackelford:** Telomerase activity exerts not just an anti-aging effect, but also a rejuvenating effect, actually reversing some aspects of the aging phenotype.

# 3D PRINTING

## ADDITIVE VS. SUBTRACTIVE

3D printers are essentially computer-controlled robots for making objects in finished or near-finished form. They work in an additive way, layer by layer, as opposed to subtractive methods that remove material by cutting, drilling, milling, etc. They also eliminate the time and expense of building molds or other specialized tooling traditionally needed to produce multiple copies. 3D printers can allow the creation of spare parts in remote locations, and can produce shapes that are difficult or impossible to create by other methods.

## EXTRUDE, FUSE, OR LAMINATE

Most often, 3D printing involves extruding a polymer that is then hardened, but some use lasers to selectively harden liquid polymers, or to fuse plastic, metal, or ceramic powder. Some work by cutting and laminating thin layers of plastic, paper, or metal.

Invented in the '80s, the first 3D printers were very expensive and mostly used for rapid prototyping, and then for producing critical but expensive objects like turbine blades. Costs have since dropped by orders of magnitude, in part thanks to the RepRap project and others in the world of open source. There are now kits and even online services that enable anyone to get started for a few hundred dollars. Just upload your design and they'll ship you the result.

## THEY'RE SQUIRTING *WHAT* THROUGH A ROBOT'S NOZZLE?

3D printing goes far beyond plastic doodads, jewelry, and even rocket engines. Recently (and controversially), working firearms have been printed in both plastic and metal. Harvard Professor Jennifer Lewis has invented special "inks" that enable printing of batteries and electrical contacts. An English company, Choc Edge, sells a chocolate printer, and others are working on printing pasta, pizza, and other foods. Living tissue can be printed, and a human liver may have been 3D printed by the time you read this. Scientists at Lawrence Livermore National Lab are now working on 3D printers that would mix different materials, such as combinations of metals and plastics, to make objects. On a larger scale, Reef Arabia is regenerating coral in the Persian Gulf using 3D printed reefs made of a sandstone material. Others are investigating means of 3D printing entire buildings.

## MANUFACTURING DEMOCRATIZED, OR THE RETURN OF THE PERSONALIZED

In pre-modern times, everything made was custom-made. The invention of interchangeable parts and mass production meant cheaper and better and many more manufactured items, but at the price of centralization and standardization. As Henry Ford famously said about the Model T, "Any customer can have a car painted any color that he wants so long as it is black."

But sometimes history is cyclic, and it looks like mass production and economies of scale aren't the last words in manufacturing. 3D printing is a technological disruption that means not just new ways of making new things, but a comeback for decentralization and personalization.

**Jay Cornell:** Organovo, a biomedical startup in San Diego, in partnership with Invetech of Melbourne, Australia, has created a 3D "bio-printer" capable of building human blood vessels, organs, and more.

Based on technology developed by Professor Gabor Forgacs of the University of Missouri, the device fits inside a standard biosafety cabinet. A computerized controller guides the print head as it deposits droplets of living tissue, using a second print head to deposit a hydrogel for support and nutrients. The cells then self-assemble. A blood vessel five centimeters long can be created in about one hour.

Any cell type can be used. The technology could potentially create replacements for lost or damaged skin, bones, cartilage, muscles, corneas, teeth, and more, and might even replace the risky and expensive process of organ transplants.

## Some Body Parts Are More Fun Than Others

**Surfdaddy Orca:** As for bigger body parts, Organovo's Dr. Forgacs thinks they may ultimately come in different shapes and sizes—designer organs. What this might mean to the sex industry . . . well, let's not go there. Organovo is definitely a company to keep an eye on.

## Printing Our Way to Post-Scarcity

**Valkyrie Ice:** In our current economy, value is determined by how "scarce" any given product is. But if your printer can print just about anything, from food to electronics to household furniture—then what can be defined as "scarce"? Right now, that printer might only be able to print using plastics and dyes and other simple materials, but it should be obvious that we will be refining those materials and enabling those printers to print ever more complex objects. As other innovations such as nanoelectronics are perfected, the complexity of both what can be printed, and what it is possible to design, will increase exponentially. Even as additive manufacturing is destroying many of the institutions of the Industrial Revolution that we have come to take for granted and even depended on for income, it will be providing ever greater access to resources and products at lower and lower costs, reducing the need for that income as we can meet our needs more cheaply. By the time the old economy of scarcity that we live in today has collapsed, additive manufacturing will have helped give rise to the economy of abundance. It's not the only development moving us toward that future, and it will not be a smooth transition by anyone's estimation, and there will be many trials and tribulations along the way. But that's been true of every major paradigm shift in human history, from the discovery of tool-making to the Industrial Revolution. Step by step, kicking and screaming in protest all the way, we still keep walking down that road toward a better future.

# TIMOTHY LEARY

*(See also Psychedelic Transhumanism)*

While Timothy Leary is best remembered as a leader of the 1960s counterculture and an advocate for the mind-expanding qualities of psychedelic drugs (and perhaps, secondarily, as someone who loved computers and the Internet), it's less well known that, in the mid-1970s, Leary was one of the few well-known people in the world to preach a transhumanist message.

## TRANSACTIONAL PSYCHOLOGY

In some ways, Leary's interest in enhancement goes back to his work with transactional psychology in the 1950s. Transactional psychology believed that humans could improve themselves, achieve greater things, and self-actualize rather than merely conform or cope (as was the view of the dominant Freudians of the time).

## IS ACID ENHANCEMENT?

When he experienced the effects of psilocybin and then LSD while a psychology professor at Harvard, Dr. Leary believed he had found a key toward helping people overcome bad "imprints" that prevented them from living to the fullest. He also believed the psychedelic substances provided novel perspectives that enhanced creativity and opened people up to new types of learning and discovery (and, of course, cosmic consciousness). And, perhaps as relevant to our current concerns, by the middle of the 1960s, Dr. Leary was just as interested in the potential for drugs and technologies that increase intelligence in the more prosaic and pragmatic sense.

## SMI²LE, DARN YA, SMI²LE

In 1974, while imprisoned for marijuana possession (and for having escaped while serving that sentence), Leary wrote (with his wife, Joanna) a monograph titled *Neurologic* that predicted a post-terrestrial humanity that would evolve into levels and types of intelligence that would not be accessible to ordinary terrestrial humans.

In 1974, after reading Gerard K. O'Neill's proposal for L5 space colonies, Leary enunciated his new direction for humanity, which he christened SMI²LE, for Space Migration, Intelligence Increase (intelligence squared) and Life Extension.

As you know by now, the potential for technologies that increase intelligence ($I^2$) and expand lifespans (LE) beyond their apparent biological limits have become core obsessions of transhumanism. The first of eight points in "The Transhumanist Declaration," (1998) reads, "We envision the possibility of broadening human potential by overcoming aging, cognitive shortcomings, involuntary suffering, and our confinement to planet Earth." In other words, SMI²LE. Leading transhumanists rarely acknowledge that Leary defined the movement with precision four decades ago.

# The Last Generation in Flesh?

**Timothy Leary and Eric Gullichsen (originally published in 1988 in *Reality Hackers*. Reprinted in *h+ magazine*):** Information-beings of the future may well be fluid. Human society has now reached a turning point in the operation of the digital programs of evolution, a point at which the next evolutionary steps of the species become apparent to us to surf as we will. Or, more correctly, as the evolutionary programs run and run, the next stages pop up in parallel, resulting in continuing explosions of unexpected diversity. Our concepts of what is known as "human" continually change. For example, we are no longer as dependent on physical fitness for survival. Our quantum appliances and improved mechanical devices can generally provide the requisite means or defenses. In the near future, the methods of information technology, molecular engineering, biotechnology, nanotechnology (atom stacking) and quantum-digital programming could make the human form a matter totally determined by individual whim, style and seasonal choice.

Humans already come in some variety of races and sizes. In comparison to what "human" might mean within the next century, we humans are at present as indistinguishable from one another as are hydrogen molecules. Along with the irrational taboo against ending death, the sanctity of our body image seems to be one of the most persistent anachronisms of Industrial Age thought.

We see evolutions of the human form in the future; one more biological-like: a bio/computer hybrid of any desired form—and one not biological at all: an "electronic entity" in the digital info-universe.

Human-AS-programs, and human-IN-programs.

Of these two posthumanist views, human-as-programs is more easily conceived. Today, we have crude prosthetic implants, artificial limbs, valves, and entire organs. The continuing improvements in the old-style mechanical technology slowly increase the thoroughness of brain/external-world integration. A profound change can come with the developments of biotechnology, genetic engineering, and the slightly more remote success of nanotechnology.

The electronic form of human-in-programs is more alien to our current conceptions of humanity. Through storage of one's belief systems as data structures online, driven by desired programs, one's neuronal apparatus should operate in silicon basically as it did on the meatware of the brain, though faster, more accurately, more self-mutably, and, if desired, immortally.

Clever cyberpunks will of course not only store themselves electronically, but do so in the form of a "computer virus," capable of traversing computer networks and of self-replicating as a guard against accidental or malicious erasure by others, or other programs. (Imagine the somewhat droll scenario: "What's on this CD?" "Ah, that's just that boring adolescent Leary. Let's go ahead and reformat it.")

Current programs do not permit matching the realtime operation speed and parallel complexity of conventional brains. But time scale of operation is subjective and irrelevant, except for the purposes of interface.

Of course, there is no reason one needs to restrict one's manifestation to a particular form. One will basically (within ever-loosening physical constraints, though perhaps inescapable economic constraints) be able to assume any desired form.

Given the ease of copying computer-stored information, it should be possible to exist simultaneously in many forms. Where the "I's" are in this situation is a matter for digital philosophers. Our belief is that consciousness would persist in each form, running independently, cloned at each branch point.

# TRANSBEMANISM

Transhumanism may seem a bit too rigid to some—what with the implicitly singular, ego-bearing human individual marching square-jawed forward into the forever future, disease-free, with great musculature and great ambition, and with a mind that presumes to calculate everything there is to know. For those who are looking for something a bit less like enhanced-more-of-the-same, and perhaps a tad more postmodern, there's transbemanism. Martine Rothblatt, who birthed the concept, describes it as "a philosophy that supports transitioning to a view of ourselves as unique patterns of thoughts (bemes), rather than as bodies per se, and consequently accepting of a 'one mind, many instantiations' society." Transbemanism advocates a fluid multiplicity of selves emerging, most likely, from an uploaded humanity, rather that the continuation of the individuated classical Western self.

## On the Beme

**Martine Rothblatt (interviewed by Roz Kaveney):** Bemes, as in "by uploading her bemes, the transhumanist was able to create a mindfile to serve as a basis for a future cyber-conscious analog of herself." The singular form, beme, refers to a digitally-inheritable unit of beingness (such as a single element of one's mannerisms, personality, recollections, feelings, beliefs, attitudes, and values).

## Floating Multiplicities

**Martine Rothblatt (interviewed by Roz Kaveney):** People don't need to be in one place, or one machine. People can exist in many places and float. People were originally disturbed by telephones, because an individual's voice could be where they are speaking and where they were heard—and now we take that for granted. A singularity of embodiment would be an obsolete concept. Just because our whole cultural matrix has been one body, one mind does not mean that this has to be where we are going. And, of course, sooner or later, different versions of the uploaded personality will have experiences different enough to make them different, though closely related, persons.

# TRANSGENDER

As transhumanism continues to be a buzzword for radical technological alteration, it awes me to recall that transgender people began getting hormone treatments in 1949. And in 1952 (the year that I, R.U. Sirius, was born) Christine Jorgensen's male-to-female gender reassignment surgery was all over the newspapers. Touted as the "first sex change surgery," Jorgensen's media coverage opened up a new world of gender change. (In fact, such surgeries were performed in Germany in the late 1920s, but by the early '50s, it was possible to get better results by combining surgery with hormone therapy.)

While it would be quite an exaggeration to claim that the West has accepted full rights for the transgendered, the fact is that the basic right of a human being to make an extraordinary transformation in the nature of his or her or hir gender was established many years ago. Sex reassignment was not stopped or shut down. Now, a woman-to-man gives birth to a baby and most of us barely bat an eye.

As we move into an age of shifting identities, where we can be whatever or whoever we choose to be in our virtual lives, where biotechnology might soon offer changes in skin melanin bringing about the age of the trans-racial, as people start to evolve novel body ornamentations and eventually parts, as we learn how to control our hormones to amp up our estrogen or testosterone to suit the needs of the day, we should always remember to thank the transgendered. They have walked point for our basic right to self-alter.

# TRANSHUMANIST TV, FILM, AND GAMES

So much of what we call science fiction features implicitly transhumanist tropes that an inclusive view would require another book. We also see a sort of transhumanism at work in all the superhero films and TV shows (superheroes are enhanced, albeit not usually by technology) and in a lot of horror. Vampires are immortal. Zombies return from "metabolic coma," although we certainly hope the folks at Alcor come out looking better than the living dead. Mutant superior (posthuman) children scare us shitless in films like *Children of the Damned*.

Games, designed to appeal to just the sort of geeks who would be hip to the upcoming mutations that may be available in the real world, often feature explicitly transhumanist themes. Steve Jackson Games published the role-playing game Transhuman Space in 2002. The story line in the Deus Ex computer game series involves an ideological conflict over augmentation. Total Annihilation, The BioShock series, and Halo are other computer games that play with transhumanist themes. Eclipse Phase, developed by Posthuman Studios, is a tabletop game that raises far-out transhuman possibilities.

There are too many movies to mention that have provided reference points for transhumanist types over the years. Among them are:

- *2001: A Space Odyssey* ("Sorry, Dave")

- *Eternal Sunshine of the Spotless Mind* (neurotechnology)

- *Ghost in the Shell* (cyberbrains and prostheses run rampant)

- *The Matrix* (even Kurzweil approves)

- *The Terminator* (inspiring scary scientists since 1984)

- *Gattaca* (best film about a bio-enhanced civilization)

- *Blade Runner* (with its rebellious replicants)

- *Brainstorm* (recording mental experience)

- *X-Men* (the superhero flicks most frequently mentioned by transhumanist sorts).

For people who like their body modification extra strange, the bio-horror films of David Cronenberg tend to picture the act of mutating as a sort-of pleasurable torture and usually involve a scientist altering humans—themselves or others—to disturbing effect.

When it comes to television, *Star Trek*, of course, was the launch code for several million geeks. Tricorder technology is now a goal of an XPRIZE contest (Scanadu has already modeled a fairly impressive one). Cloaking devices are starting to appear (or should we say disappear), although none that would hide a spacecraft.

*Max Headroom*, an AI personality in cyberspace, helped launched the cyber era in the US in 1987, although a better version of the program was introduced in Great Britain in 1985. The SF series *Battlestar Galactica*—and its unfortunately short-lived prequel, *Caprica*, were favored by transhumanists, with its cybernetic race indistinguishable from humans. The current (and awesome) BBC America series *Orphan Black* revolves around a cloned woman and a hilarious bad guy named Dr. Aldous Leekie (because Leary would have been too obvious?), whose Neolution movement essentially *is* transhumanist. Leekie is played by Matt Frewer, who also played Max Headroom.

By now, television is swamped with technologically enhanced humans fighting crime and other technologically enhanced humans. In one night, I (R.U. Sirius) watched three shows in a row on network TV that revolved around enhancement implants, advanced AIs, or both—*Marvel's Agents of S.H.I.E.L.D*, *Intelligence,* and *Person of Interest.*

The presence of transhumanist themes in entertainment is so ubiquitous that we can only apologize for failing to note your favorite.

## Eclipse Phase Is Singularity Institute Crossed with Lifeboat Foundation

**Ray Huling:** Posthuman Studios game developers Rob Boyle and Brian Cross set out to contend with the broadest variety of transhuman notions.

> Eclipse phase is a biological term, referring to the moment between a virus's infection of a cell and its duplication within it. The game occurs at a time when the full spectrum of transhumanity lies scattered across the stars. Humanity has divided into factions, differentiated by their socio-political practices and by their distance from baseline humanity. Vastly powerful artificial intelligences lurk in the darkness and a mysterious plague, the Exurgent Virus, has wrought deep transformations on swaths of transhuman civilization.

Eclipse Phase enlists players as agents of Firewall, a conspiracy that crosses the transhuman factions. "If you took The Singularity Institute and the Life-boat Foundation and meshed them together in this setting . . . and put them underground," says Boyle, "That's kind of what this Firewall organization is."

## Battlestar Galactica

**Lauren Davis:** The 2004–2009 post-apocalyptic series on Sci-Fi Channel (now SyFy) about humans on the run after their homeworlds are destroyed by the Cylons—mechanical and biological androids created by humans—addressed sociopolitical issues aplenty: warfare, religion, classism, the role of the democratic process in times of crisis, and the paranoia that results when the enemy looks just like you. But the Cylons are more than a metaphor for terrorism. Their ability to download themselves into fresh bodies allows them to experience and learn from death. And since most Cylon models exist as multiple copies with the same appearance and base programming, many humans question their capacity to act as individuals. Finally, while some Cylons seek to bridge the gap between man and machine, others resent and resist taking on human qualities, believing biology is their only obstacle to transcendence.

## Better Off Ted: A Bit Ahead of Its Time, Perhaps?

**Lauren Davis:** In this corporate satire (ABC, 2009–2010), Ted Crisp (Jay Harrington) headed up the research and development department at international megacorporation Veridian Dynamics, acting as a liaison between Veridian's brilliant but socially awkward researchers and its shadowy executives. Need to cryogenically freeze one of your employees? Grow meat in a lab? Weaponize a pumpkin? At Veridian Dynamics, no innovation is impossible.

## We're All Neural Dolls

**Erik Davis:** *Dollhouse* (2009–2010, Fox) took a reasonably meaty bite out of one of the more ominous and potentially liberating conundrums of 21st century life: the thoroughly constructed nature of human identity. The show framed this conundrum in terms of neuroscience and the pervasive pop metaphor of the mind as a programmable input-output device. Original personalities are "wiped" and stored on cartridges that resemble old 8-track tapes; other "imprints" are not only shuffled between the dolls but remixed into the perfect blend of characteristics for any given job. The show's ambivalence about such "posthuman" technologies was captured by the character who does all the wiping and remixing: a smug, immature, and charmingly nerdish wetware genius named Topher Brink, whose simultaneously dopey and snarky incarnation by the actor Fran Kranz reflected the weird mix of arrogance and creative exuberance that informs so much manipulative neuroscience.

*Dollhouse* asked, what if we are just a cluster of neurons? And what does that possibility do to our understanding of morality and choice, fantasy and personality?

## AfroCyberPunk

**Jonathan Dotse:** Imagine a young African boy staring wide-eyed at the grainy images of an old television set tuned to a VHF channel; a child discovering for the first time the sights and sounds of a wonderfully weird world beyond city limits. This is one of my earliest memories growing up during the mid-nineties in a tranquil compound house in Maamobi, an enclave of the Nima suburb, one of the most notorious slums in Accra. Besides the government-run Ghana Broadcasting Corporation, only two other television stations operated in the country at the time, and satellite television was way beyond my family's means. Nevertheless, all kinds of interesting programming from around the world occasionally found its way onto those public broadcasts. This was how I first met science fiction; not from the tomes of great authors, but from distilled approximations of their grand visions.

This was at a time when cyberpunk was arguably at its peak, and concepts like robotics, virtual reality, and artificial intelligence were rife in mainstream media. Not only were these programs incredibly fun to watch, the ideas that they propagated left a lasting impression on my young mind for years to come. This early exposure to high technology sent me scavenging through piles of discarded mechanical parts in our backyard; searching for the most intriguing sculptures of steel from which I would dream up schematics for contraptions that would change the world as we knew it. With the television set for inspiration and the junkyard for experimentation, I spent my early childhood immersed in a discordant reality where dreams caked with rust and choked with weeds came alive in a not-so-distant future.

I am only now able to appreciate the significance of this early exposure to high technology in shaping my outlook on the world. From my infancy I became keenly aware of the potential for science and technology to radically transform my environment, and I knew instinctively that society was destined to continue being reshaped and restructured for the rest of my life. Mind you, I am only one of many millions in a generation of African children born during the rise of the global media nation. Children raised on Nigerian movies and kung fu flicks; Hindi musicals and gangster rap; Transformers, Spider-Man, and Ananse stories; BBC, RFI, and Deutsche-Welle TV; the Nintendo/Playstation generation. Those of us born in this time would grow up to accept the fact that the only constant was change, that the world around us was perceptibly advancing at an alarming pace, that nothing would ever remain the same.

# VIRTUAL REALITY

Way back at the start of the '90s, people at the edge of the emerging digital culture talked about virtual reality (VR)—the idea that we would soon interact in shared 3D worlds—as much as, if not more than, they talked about the Internet. (Of course, we "early adopters" were often talking about it *on* the Internet, so I guess sometimes you just don't notice your immediate surroundings.) These 3D worlds would be accessed through head-mounted displays. The idea was to put the user literally inside computer-created worlds, where she could move around and have the sensation of being in another world. The eyes were the primary organs of entrance into these other worlds, although sound and motion were involved and touch was said to be a work in progress. VR would be a type of entertainment, sure, but more than that, it would be an extremely creative interactive communications medium. You'd go online and interact with others in worlds of your individual or mutual creation.

There were dozens of conferences about VR and lots of national media coverage in every major outlet. There were movies and TV shows that revolved around VR, and there was even one arcade game. But the hope for the spread of this type of virtuality faded. The VR systems were simply not good enough yet to create a consumer market—data processing capacity being a big part of the problem.

## PRACTICAL BORING STUFF THAT MATTERS

VR didn't go away. Its use became fairly ubiquitous in the world of architectural design, for oil and gas exploration, and for designing novel drugs.

More recently, it has proven useful for treating anxiety. You can experience, for example, looking over the edge of a tall building knowing that you are fundamentally safe, until your vertigo recedes. It's also been used in improved flight and driving simulators.

## GET A SECOND LIFE

And then there's Second Life. While the popular virtual world doesn't put you inside a 3D computer reality, it does many of the things that VR promised. It lets you be a character (or thing) in a self or other created world online and interact with other people (or avatars) there. It allows users to create visually available realistic or fantasy environments, including ones that defy the laws of physics. And it lets people form virtual cultures—cultures that exist only online. While it's essentially an extension of online role playing games—an MMORPG with moving visuals—many people using Second Life have found it compelling enough that their Second Life avatars have taken on a life of their own. Some are even demanding some rights, like the right to sign real-world contracts under their virtual names.

## NOW, ONLY TWO DECADES LATE . . .
## *IT'S CONSUMER VR!*

Meanwhile, we've been moving back toward consumer VR. It started with gaming hardware like the Nintendo Wii. The Wii took a concept from early VR research, the data glove, and built a controller that allows you to move things inside the computer screen.

Finally, we have the imminent arrival of Oculus Rift. Oculus Rift is the big consumer VR breakthrough. While we have seen the release of a few head-mounted displays for gaming since the middle of the '90s, the equipment was hard to wear and the head tracking was glitchy, making for an unrealistic response to attempts to move through 3D worlds. With Oculus Rift, we finally have a head-mounted display that isn't horribly heavy and cumbersome. Rift comes with earphones attached so users are set for sound as well as vision. Its high-resolution head tracking sensors allow people to move intuitively inside 3D worlds and have them respond appropriately.

Already generating pretty much universal rave reviews from people who have been given a sneak preview, Oculus Rift is expected to hit the consumer market by early 2015. By 2016, millions of people will likely be building and interacting in virtual worlds. And only a little over two decades later than we all predicted back in '90s!

## Virtual People Experience Real World Discrimination

**Stephen Euin Cobb:** "I must have lost almost half of my potential contracts because the companies wouldn't deal with an anonymous avatar." So says Scope Cleaver, a designer and architect inside Second Life. Praised by *New York Times Magazine* for his design of Princeton University's Diversity Building (the article headline: "Architectural Wonders of the Virtual World"), his creations have extended his reputation beyond Second Life and across several continents, but even that can't protect him from what appears to be discrimination. "I offered the companies a real world proxy who could sign all the papers, but it didn't seem to help."

Some people see the freedom of anonymity that virtual worlds give them as a nice perk. Others enter virtual worlds to promote their real world selves, or projects, and avoid anonymity for their avatars as much as possible. But for thousands, keeping their avatar's identity separate from their real world identity is a serious philosophic matter. They believe they should strive to be the people they are in their hearts and minds, rather than the person suggested by features of their physical body that are observable on the outside. After all, these external features were forced on them. Ethnicity is the cliché example, but other accidents of birth that either can't be changed—or can't be easily changed—include age, gender, stature, attractiveness, nationality, social class, the accent of their birth language, even regional dialect. None of these were chosen, and they are impossible or difficult to change in the physical world. Calling themselves Digital People, they design avatars that better fit their self-image, and then use them to build reputations, personalities, and social circles that also better fit them.

Those who oppose this philosophy feel that Digital People present a false self to the world—a grand and elaborate lie. Bad feeling has accumulated as the result of social pressure and insults experienced by Digital People. Even non-Digital People who mean well have shown remarkable intolerance.

Digital People who rely less on non-digital people tend to experience something more akin to confusion than discrimination. Extropia DaSilva (a Digital Person who is also a transhumanism activist, essayist, and text-based public speaker) explained, "It is not uncommon for people to ask out loud if I have Multiple Personality Disorder after I explain what a digital person is."

## Pope Benedict XVI Warns That the Virtual World May Obscure Truth

**Chris Arkenberg:** The Pope warned about the dangers of an increasingly virtualized world, arguing that new media technologies and the proliferation of images threaten the search for truth. Benedict XVI cautioned that the image can "become independent of reality; it can give life to a virtual world, with several consequences, the first of which is the risk of indifference to truth."

Benedict's warning is itself paradoxical and warrants some contextual analysis. The Papacy has advanced its own version of the biblical virtual world for millennia suggesting that its allegorical and mythological stories be taken as truth. This evangelism does not so much advocate "indifference to truth" but it works to supplant contrary truths with its own version. Indeed, any visit to a Catholic church will present one with a vivid and compelling array of imagery tailored to the particular brand of truth they proselytize. In effect, these set dressings are visual entry points—hyperlinks—to the virtual world of the Catholic faith.

# WARBOTS

We don't like war, but we're endlessly fascinated by it, particularly when there's extreme technology involved. The notion of the "pushbutton" automated war has been in circulation in the US at least since the Vietnam War, but it seems as though we may have reached the proverbial tipping point with the second Iraq War, and then the emphasis on drone warfare under the Obama administration. And now that bots designed primarily for the military have been bought up by "Don't be evil" Google, we could see accelerating change of a dangerous sort.

## CLEAN EFFICIENT WAR OR EFFICIENT UNSTOPPABLE SLAUGHTER?

Then again, intelligent machines could make war efficient in an arguably good way, by making actions more precise, reducing, or (dare we hope) even eliminating civilian casualties ("collateral damage"). But there again, with the much-ballyhooed exponential growth in technopower, these things might get so much cheaper and easier to make (or grow), they may spread beyond control. The Joker (or someone who thinks he's The Joker) might send killer bots to shoot up the neighborhood mall and then march them down the highway, then law enforcement sends *its* killer bots, and you find yourself living in a bad science fiction movie.

## TODAY'S MILITARY BOTS

Today, robots are used for hazardous tasks such as defusing bombs or mines. The military uses thousands of drones for reconnaissance and attack. Northrop Grumman's unmanned X-47B has taken off from and landed on an aircraft carrier without any human guidance.

TRANSCENDENCE

Boston Dynamics, a major player in the field now owned by Google, is developing various robots with DARPA support. Their four-legged LS3 and BigDog robots can travel pretty much anywhere a mule can, while carrying hundreds of pounds of cargo. Experiments have been made with tracked robots as gun platforms. In response, the UN has called for a moratorium on the development of lethal autonomous robots, though it's unlikely that will even slow development. In fact, in January 2014, US General Robert Cone suggested that the US could soon be replacing thousands of soldiers with robots.

## DO YOU TRUST THE GOVERNMENT WITH WARBOTS?

So we have to worry about terrorists and criminals, right? Well, there are some people who also aren't all that comfortable with trusting the agendas of even our most democratic states. So where does that leave us? *Worried.*

### From Zero Warbots on the Ground to 12,000

**Surfdaddy Orca:** There has been a dramatic increase in the use of ground robotics. When US forces went into Iraq in 2003, they had zero robotic units on the ground. In 2009, there were as many as 12,000.

Some of the robots were used to dismantle landmines and roadside bombs, but a new generation of bots is designed to be fighting machines. One bot, known as SWORDS, can operate an M16 rifle and a rocket launcher.

In the 2009 film, *Terminator Salvation*, the fictional Skynet computer network directs a variety of hunter killer robots: aerial- and land-based drones, as well as motorcycle-like Mototerminators, serpent-shaped Hydrobots, and the terrifying and gigantic Harvesters.

Alarmingly, many of these bots exist in some form today—drones like Predator and Reaper, the ground-based TALON, and iRobot's PacBots and BigDogs.

## When a Warbot Went Haywire

**Surfdaddy Orca:** What happens when robots decide what to do on their own? One nightmare real-life incident was reported in the *Daily Mail*.

"There was nowhere to hide," one witness stated. "The rogue gun began firing wildly, spraying high explosive shells at a rate of 550 a minute, swinging around through 360 degrees like a high-pressure hose."

A young female officer rushed forward to try to shut the robotic gun down—but it was too late. "She couldn't, because the computer gremlin had taken over," a witness later said.

The rounds from the automated gun ripped into her and she collapsed to the ground. By the time the robot had emptied its magazine, nine soldiers lay dead (including the woman officer). Another 14 were seriously injured. A government report later blamed the bloodbath on a "software glitch."

The robotic weapon was a computer-controlled MK5 anti-aircraft system, with two huge 35mm cannons. The South African troops never knew what hit them.

## Could Warbots Be More Humane?

**Peter Asaro (interviewed by R.U. Sirius and Surfdaddy Orca):** I think we are now starting to see robots that are capable of taking morally significant actions, and we're beginning to see the design of systems that choose these actions based on moral reasoning. In this sense, they are moral, but not really autonomous because they are not coming up with the morality themselves . . . or for themselves. We might be able to design robotic soldiers that could be more ethical than human soldiers.

Robots might be better at distinguishing civilians from combatants, or at choosing targets with lower risk of collateral damage, or understanding the implications of their actions. Or they might even be programmed with cultural or linguistic knowledge that is impractical to train every human soldier to understand.

Ron Arkin thinks we can design machines like this. He also thinks that because robots can be programmed to be more inclined to self-sacrifice, they will also be able to avoid making overly hasty decisions without enough information. Ron also designed architecture for robots to override their orders when they see them as being in conflict with humanitarian laws or the rules of engagement. I think this is possible in principle, but only if we really invest time and effort into ensuring that robots really do act this way.

So the question is how to get the military to do this. It does seem like a hard sell to convince the military to build robots that might disobey orders.

## Terminator? Sounds Awesome, Dude!

**P.W. Singer (interviewed by R.U. Sirius):** One of the scientists building war-bots talked with incredible admiration about the robots in the opening scene of *Terminator 2*, where the robots are walking across the battlefield. He was like: "This is incredibly impressive stuff." You know, yeah, it's stepping on a human skull, but it's still really impressive.

Another scientist talked about how the military came to him and said, "Oh, we'd like you to design the hunter-killer drone from the Terminator movies." Which, you know, is kind of incredibly scary, but it makes perfect sense from another perspective in that if it's effective for SkyNet, their thinking is, "Well, it could be really neat in our real-world battlefields."

I remember doing one of the interviews with an Air Force officer and asking him, "What do you think the experience of a predator drone attack is like?"

And he said, "You know, it's probably like the opening scene of the Terminator movies, where the humans are hiding out in the caves and the bunkers, and this sort of relentless robotic foe is coming at them. That's what I bet it's like for the Al-Qaeda and Taliban." And there's sort of an irony there, in that we—the watchers of the movie—are supposed to be cheering for the humans on the ground. And the humans on the ground eventually get over their fears of this relentless foe and fight back. So it's a sort of weird irony when you think about it that way.

# Refuseniks

**P.W. Singer (interviewed by R.U. Sirius):** The refuseniks are scientists working in robotics who are starting to worry that they are becoming too much like the Dyson character in the Terminator chronology. Dyson was the scientist who invented SkyNet and then learned to his horror what it did. And he has this wonderful quote: "You're judging me on things I haven't even done yet. Jesus. How were we supposed to know?" And that's in the world of fiction.

The refuseniks are also thinking about the world of history, most particularly what happened in the 1940s to the nuclear physicists who became so enamored of building this incredible technology—what became the atomic bomb—that they never took a step back and went, "Oh my gosh. What does this all mean?" And even more to the point, they tricked themselves into believing they were going to be the ones in charge of how it would be used. And, of course, after the Manhattan Project, they weren't the ones in charge. And a lot of them asked themselves, "My God, what did I do?" And a lot of the people who invented the atomic bomb then became the founders of the arms control movement to ban the atomic bomb. So the refuseniks are real-world roboticists . . . a sort of small movement in the robotics field. As I joke, they're the roboticists who just say no. They're the ones who are saying, "You know what? I don't want to accept military funding for what I'm doing. I'm not going to work on warbots."

# XPRIZE

*(See also Gamification)*

The successful brainchild of Peter Diamandis, the XPRIZE started out as a plot to jumpstart the civilian commercial space industry. In 1996, Diamandis organized the "Ansari XPRIZE," which offered a ten million dollar reward for the first independent (not government supported) group that could launch a spacecraft one hundred kilometers about the Earth's surface twice within two weeks. The prize was won in 2004 by aerospace designer Burt Rutan and Paul Allen of Microsoft fame.

Since then, the XPRIZE Foundation has offered rewards for innumerable challenges. Among those that have been won was a challenge to build a vehicle that gets one hundred miles to the gallon and has mass market potential, a challenge to design a technology that can speed up oil cleanup by a factor of three, and a contest to build small and efficient rocket systems.

Among the challenges still in competition, one involves landing a rover on the Moon and another involves creating the equivalent (more or less) of the *Star Trek* tricorder. The winning technology has to be able to diagnose a patient "equal to or better than a panel of board certified physicians."

The XPRIZE has proven very successful so far, not so much because the monetary reward is so high (some people are spending more than they're making back), but because people like these sorts of clear-cut competitions. As such, it's a great example of gamification.

# ZERO STATE

Zero State is an organization that advocates "positive social change through technology." Its goal is to establish "a trans-national virtual state." While Zero State emphasizes life enhancement, cognitive enhancement, and accelerated change, it is fairly outside the transhumanist movement's mainstream.

The organization also advocates for basic income and for meshnets, a type of system in which every node (every online presence) relays data for the entire network. Meshnets allow messages to be sent directly from person to person through a dedicated line, making it harder to surveil such communications.

Zero State is also dedicated to the notion of Mutual Aid, a concept from the Russian "left anarchist" Peter Kropotkin. Zero State, however, emphatically rejects these labels.

British cognitive science researcher Amon Kalkin founded Zero State and he is still, unofficially, the primary spokesman for the group.

# ADDENDUM

Transhumanist breakthroughs continue at an accelerating pace. Here are some developments that have happened since this book was written, but before it went to press.

- Facebook announced its intention to purchase Oculus VR, Inc., makers of the virtual reality headset Oculus Rift.

- Neurosurgeons at University Medical Center Utrecht in the Netherlands replaced the upper part of the skull of a young woman who suffered from bone disorder with a 3D printed skull. The operation saved her life and fixed her vision problems and headaches.

- Doctors at UPMC Presbyterian Hospital in Pittsburgh will place ten patients with life-threatening gunshot or knife wounds in suspended animation, theoretically allowing them more time to fix the injuries. The lead surgeon refuses to use the term "suspended animation."

- A team of Stanford researchers have discovered a means of turning graphite into a diamond-like film without the extremely high pressures normally used.

- Researchers in Seoul, South Korea, and the University of Texas at Austin in the US have developed an electronic patch—euphemistically called "electronic skin"—that comes equipped with electronic sensors and memory that delivers drugs into the bloodstream.

- Duke University scientists have used stem cells to create "fake" muscles that heal themselves and could be used to repair muscle injuries.

- Researchers at the Allen Institute for Brain Science published the most sophisticated wiring diagram of a mammalian brain (a mouse) thus far.

- DARPA announced a four-year initiative to create a brain implant that can restore memories.

- A living "mutant" microbe that can carry and pass down a genetic code of six base pairs (instead of the usual four) was created by scientists at Scripps Research Institute in California. This is a staggering development, both in terms of how scientists understand life and in its potential use for nanotechnology and other major technological breakthroughs.

- Sony announced Morpheus, a virtual reality headset for PlayStation 4.

- DARPA announced a five-year project to test brain implants that can fight mental disorders.

- University of California at San Diego scientists have made and erased memories in rats using optogenetics.

- Researchers at Duke University found a type of neuron that can tell stem cells to make more new neurons. The scientists hope that they will find ways to "engage certain circuits of the brain to lead to a hardware upgrade."

- Researchers at Lawrence Berkeley National Laboratory have built a field-effect transistor (FET) from one-atom-thick layers of graphene, boron nitride, and molybdenite, creating the first two-dimensional FET. This could lead to higher-speed electronics.

- Using the same techniques used to create living animal clones, scientists in South Korea and the United States are cloning two humans to produce embryos for the production of stem cells, but are stopping short of allowing them to grow into fully cloned human beings.

- In a contest conducted by the Royal Society in London, a computer program convinced 33 percent of the judges that it was a real thirteen-year-old Ukrainian boy, causing some to claim that it had passed the Turing Test. Most AI researchers either disagree with that conclusion or call the Turing Test a bunch of crap.

- Elon Musk opened up the patents for Tesla's electric cars. Everybody swooned.

- Scientists at the Research Institute of Molecular Pathology in Vienna, Austria, have used optogenetics to film the simultaneous activity of all 302 neurons in the brain of a nematode worm.

- Researchers at Stanford have invented a method for creating large numbers of cardiomyocytes (heart muscle cells) from induced pluripotent stem cells.

- A Boston University engineer has created a new type of artificial pancreas, controlled by a modified iPhone, that may help control Type 1 diabetes.

# RECOMMENDED READING

**NONFICTION**

*The Transhumanist Reader: Classical and Contemporary Essays on the
Science, Technology, and Philosophy of the Human Future* by Max
More and Natasha Vita-More, Wiley-Blackwell, 2013

*Abundance: The Future Is Better Than You Think* by Peter Diamandis and
Steven Kotler, Free Press, 2012

*A Cosmist Manifesto: Practical Philosophy for the Posthuman Age* by Ben
Goertzel, Humanity+, 2010

*Ending Aging: The Rejuvenation Breakthroughs That Could Reverse
Human Aging in Our Lifetime* by Aubrey de Grey, St. Martin's
Press, 2007

*The Singularity Is Near: When Humans Transcend Biology* by Ray
Kurzweil, Penguin Books, 2006

*Radical Evolution: The Promise and Peril of Enhancing Our Minds,
Our Bodies—and What It Means to Be Human* by Joel Garreau,
Broadway Books, 2006

*More Than Human: Embracing the Promise of Biological Enhancement* by
Ramez Naam, Broadway Books, 2006

*Last Flesh: Life in the Transhuman Era* by Christopher Dewdney,
HarperCollins Canada, 1998

*Exo-Psychology: A Manual on the Use of the Human Nervous System
According to the Instructions of the Manufacturers* by Timothy Leary,
Starseed/Peace Press, 1977

*Up-Wingers: A Futurist Manifesto* by F.M. Esfandiary, John Day
Company, 1973

## FICTION

*Rainbows End* by Vernor Vinge, Tor Books, 2006

*Marooned in Realtime* by Vernor Vinge, Blue Jay Books/St. Martin's Press, 1986

*A Fire Upon the Deep* by Vernor Vinge, Tor Books, 1992

*Accelerando* by Charles Stross, Ace, 2006

*Singularity Sky* by Charles Stross, Ace Books/Berkeley Publishing, 2004

*The Diamond Age* by Neal Stephenson, Bantam Spectra, 1995

*Schismatrix* by Bruce Sterling, Arbor House, 1985

*Blood Music* by Greg Bear, Arbor House, 1983

*Methuselah's Children* by Robert Heinlein, Gnome Press, 1958 (Originally serialized in *Astounding Science Fiction,* 1941)


## WEBSITES

H+ Magazine
  http://hplusmagazine.com/
Singularity Hub
  http://singularityhub.com/
Next Big Future
  http://nextbigfuture.com/
Kurzweil Accelerating Intelligence
  http://www.kurzweilai.net/

# ABOUT THE AUTHORS

**R.U. Sirius (Ken Goffman)** is a writer, editor, and well-known digital iconoclast. He was co-publisher of the first popular digital culture magazine, *MONDO 2000*, from 1989–1993 and co-editor of the popular book *MONDO 2000: A User's Guide to the New Edge*. He has written about technology and culture for *Wired*, *Rolling Stone*, and *Boing Boing*. He also lectures widely.

**Jay Cornell** is the former managing editor of *h+ Magazine* and senior web developer at Landkamer Partners.